NANO-OPTICS AND NANOPHOTONICS

NANO-OPTICS AND NANOPHOTONICS

The Springer Series in Nano-Optics and Nanophotonics provides an expanding selection of research monographs in the area of nano-optics and nanophotonics, science- and technology-based on optical interactions of matter in the nanoscale and related topics of contemporary interest. With this broad coverage of topics, the series is of use to all research scientists, engineers and graduate students who need up-to-date reference books. The editors encourage prospective authors to correspond with them in advance of submitting a manuscript. Submission of manuscripts should be made to the editor-in-chief, one of the editors or to Springer.

Please view available titles in *Nano-Optics and Nanophotonics*
on series homepage http://www.springer.com/series/8765

Takashi Yatsui

Nanophotonic Fabrication

Self-Assembly and Deposition Techniques

With 104 Figures

 Springer

Takashi Yatsui
University of Tokyo
Department of Electrical Engineering, Information Systems
Bldg. 9, Yayoi 2-11-16, 113-8656 Tokyo, Japan

Nano-Optics and Nanophotonics   ISSN 2192-1970   e-ISSN 2192-1989
ISBN 978-3-642-43842-4   ISBN 978-3-642-24172-7 (eBook)
DOI 10.1007/978-3-642-24172-7
Springer Heidelberg Dordrecht London New York

Printed on acid-free paper

Springer is part of Springer Science+Business Media (www.springer.com)

Preface

This book outlines "nanophotonic" fabrication, a subject in which the author was engaged for 15 years. The author joined this field in the mid-1990s, and the idea of nanophotonics was proposed by Professor Ohtsu in 1993. The optical near-field was used as a light source for high-resolution microscopy beyond the diffraction limit of light. With the highly reproducible fabrication technique of fiber probes, the near-field microscopy can obtain optical images on the scale of a single nanometer in spatial resolution. Further progress was achieved by developing high-throughput optical fiber probes. Thus, the optical near-field can be applied to optical fabrication, including ultra-high-density optical memory, photolithography, and single-nanometer-scale photochemical vapor deposition (PCVD).

In addition to the advantage of exceeding the diffraction limit of light using the optical near-field (i.e., quantitative innovation), this achievement demonstrated that exceeding the diffraction limit of light is not essential but is simply a secondary aspect of nanophotonics. The fundamental feature "nanophotonics" is the novel functions and phenomena and originate from the intrinsic optical near-field interaction. The optical near-field interaction uses a virtual photon and can therefore be localized on the nanoscale. Additionally, the virtual photon interaction can utilize optically "inactive" states, including multiple phonon states in a material. These features of nanophotonics represent a qualitative innovation in optical science and technology because they could not be realized using conventional propagating light.

This book focuses on "nanophotonic" fabrication. With the qualitative innovation of nanophotonics, optical fabrication can be achieved on the subnanometer scale. Although there are some nanoscale fabrication techniques that use top-down lithography, it is not clear whether the fabricated structures are functional. Because the size of the fabricated devices is on the nanoscale, they are operated by the optical near-field. Thus, by using the optical near-field during fabrication, a structure is fabricated automatically so that it exhibits the best performance, i.e., the nanophotonic fabrication gives rise to new functions in the fabricated structure. The author intends for this book to be an introduction for readers, who will be interested in various functions that differ depending on the device.

The author wishes to express special thanks to Dr. M. Ohtsu, Editor of the Springer Series in Optical Sciences, and Professor at The University of Tokyo for recommending this publication. He also thanks Drs. T. Kawazoe, W. Nomura, K. Kitamura, and N. Tate (The University of Tokyo); Dr. M. Naruse (National Institute of Information and Communications Technology); Professors I. Banno, K. Kobayashi, and H. Hori (Yamanashi University); Professors M. Sugiyama, H. Fujita, and M. Washizu (The University of Tokyo); Professor M. Yoshimoto (Tokyo Institute of Technology); Professor G.-C. Yi (Seoul National University), Mr. K. Hirata, Dr. T. Morimoto, and Mr. Y. Tabata (SIGMA KOKI Co., Ltd.), Dr. Y. Yanase, Mrs. K. Suzuki, M. Fujita, A. Kamata (Covalent Materials Corporation); Mrs. K. Ito and H. Kawamura (Nitto Optical Co., Ltd.); Dr. M. Mizumura (V-Technology Co., Ltd.); and graduate students of Professor Ohtsu's research group for their active support and discussions. This work was partially supported by the New Energy and Industrial Technology Development Organization (NEDO) and Japan Society for the Promotion of Science(JSPS). Finally, Dr. Claus Ascheron of Springer-Verlag is acknowledged for his guidance and suggestions.

Tokyo   *Takashi Yatsui*

Contents

Chapter 1
Introduction

1.1 Utilization of the Optical Near-Field

Progress in DRAM technology requires improved lithography. It is estimated that technology nodes should reach to 16 nm by the year 2019 [1]. Recent improvements in the immersion lithography using an excimer laser (wavelengths of 193 nm and 157 nm) have resulted in the technology nodes as small as 45 nm. Further decreases in the node size are expected using an extreme ultraviolet (EUV) light source with a wavelength of 13.5 nm. However, the resolution of the linewidth is limited by the diffraction limit of the light. To overcome this limitation, it is necessary to use the optical near-field to exceed the diffraction limit of light. The concept of the optical near-field was proposed [2–4] as a localized electric field at a metallic aperture (Fig. 1.1). Based on scalar theory [5], when the size of the aperture is much smaller than the wavelength, the scattered light has a larger wavenumber than the incident light. Additionally, to hold the law of momentum conservation, the wavenumber normal to the aperture has a negative value. That is, the scattered light normal to the aperture cannot propagate. Thus, it is referred to as an evanescent wave. Over the past decade, utilization of the optical near-field has led to significant progress in various applications, including nanoscale imaging [6, 7], ultra-high density recording [8, 9], and plasmonics [10].

The optical near-field was commercialized due to improvements in fiber probes. Although the generation of the optical near-field was predicted in the 1920s [2], it was not realized until the fiber probe was developed as an optical near-field source. Initial advancements were made in the field of scanning probe microscopy using the pulling/etching technique [11]. Reproducible high-throughput fiber proves were developed using the sophisticated method of chemical etching process, the reproducible high throughput fiber probes were developed [12, 13]. More recently, the author, Yatsui, developed a fiber probe with a throughput of 0.1 for an aperture diameter of 100 nm [14–17] (Fig. 1.2).

In addition to generating optical near-field localization, another unique property was developed. In the nanoscale materials, a photon can be coupled with an electron,

T. Yatsui, *Nanophotonic Fabrication*, Nano-Optics and Nanophotonics,
DOI 10.1007/978-3-642-24172-7_1, © Springer-Verlag Berlin Heidelberg 2012

Fig. 1.1 Schematic of the generation of the optical near-field at the nanometer scale aperture

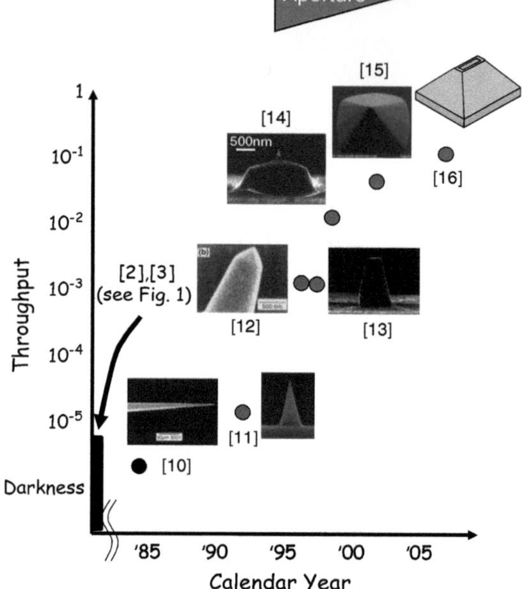

Fig. 1.2 Progress in the fiber probe

Fig. 1.3 Schematic of dressed photon and phonon

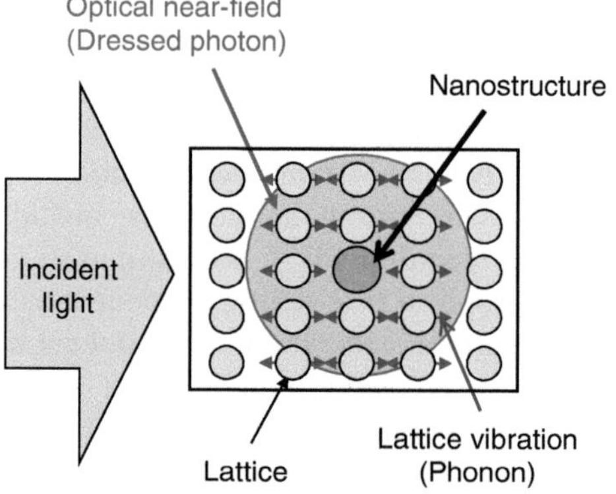

known as a dressed photon. The dressed photon is a virtual photon, and it can also couple with a phonon as a lattice vibration in the surroundings (Fig. 1.3). In their final form, the dressed photon and phonon act as a quasipatrticle, in which the energy of the dressed photon and phonon, E_{DP-P}, is larger than that of the

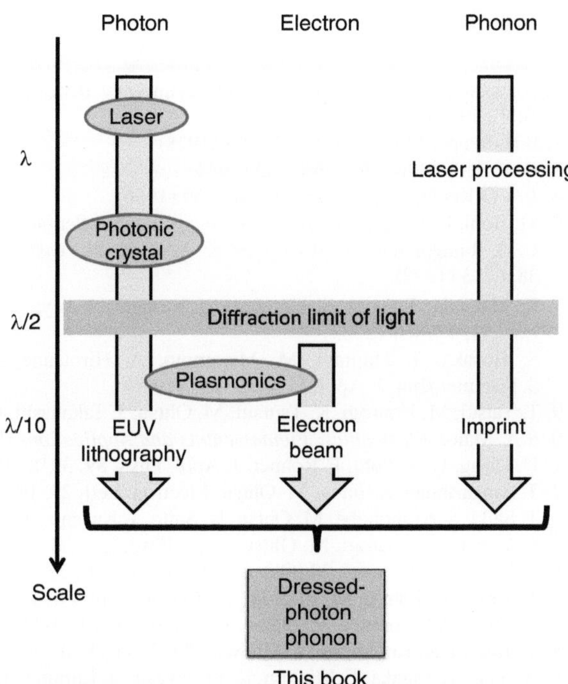

Fig. 1.4 Fabrication trend using various carriers

dressed photon, E_{DP}, and the free photon, E_{FP} ($<E_{DP}$) [18, 19]. This indicates that at the nanoscale, photons can be excited, and this can be achieved using a unique fabrication process.

Various nanoscale fabrication processes have been developed using phonons (laser processing [20] and imprint [21]), electrons (electron beam (EB) lithography [22]), and photons (photolithography [23]). In the nanoscale materials, phonons, electrons, and photons can be coupled with each other and they cannot be treated individually. The coupled state of these three carriers is known as the dressed photon and phonon (see Fig. 1.4). In this book, we treat the photons as being in the mixed state of a phonon, electron, and photon, i.e., a dressed photon and phonon.

Before discussing the dressed photon and phonon, this book begins by addressing the feasibility of nanoscale chemical vapor deposition using optical near-field techniques, which can achieve size and position control on the nanometer scale is reviewed in Sect. 1.2. A probe-less fabrication method for mass-production is also demonstrated (see Sect. 1.3). Based on the near-field fabrication technique, nanophotonic fabrication using the dressed photon and phonon is described in Sect. 1.4.

References

1. For example, see the International Technology Roadmap for Semiconductors (2011) http:// public.itrs.net/
2. E.H. Synge, Phil. Mag. Ser. **7**, 356 (1928)
3. H.A. Bethe, Phys. Rev. **66**, 163 (1944)
4. J.A. O'Keefe, J. Opt. Soc. Am. **46**, 359 (1956)
5. M. Born, E. Wolf, *Principles of Optics*, 6th edn. (Pergamon, Oxford, 1980)
6. U.M. Rajagopalan, S. Mononobe, K. Yoshida, M. Yoshimoto, M. Ohtsu, Jpn. J. Appl. Phys. **38**, 6713 (1999)
7. K. Matsuda, T. Saiki, S. Nomura, M. Mihara, Y. Aoyagi, S. Nair, T. Takagahara, Phys. Rev. Lett. **91**, 177401 (2003)
8. S. Hosaka, T. Shintani, M. Miyamoto, A. Hirotsune, M. Terao, M. Yoshida, K. Fujita, S. Kämmer, Jpn. J. Appl. Lett. **35**, 443 (1996)
9. T. Yatsui, M. Kourogi, K. Tsutsui, M. Ohtsu, J. Takahashi, Opt. Lett. **25**, 1279 (2000)
10. S.A. Maier, *Plasmonics: Fundamentals and Applications* (Springer, 2007)
11. U. Dürig, D.W. Pohl, F. Rohner, J. Appl. Phys. **59**, 3318 (1986)
12. T. Pangaribuan, S. Jiang, M .Ohtsu, Electron. Lett. **29**, 1978 (1993)
13. T. Saiki, S. Mononobe, M. Ohtsu, N. Saito, J. Kusano, Appl. Phys. Lett. **68**, 2612 (1996)
14. T. Yatsui, M. Kourogi, M. Ohtsu, Appl. Phys. Lett. **71**, 1756 (1997)
15. T. Yatsui, M. Kourogi, M. Ohtsu, Appl. Phys. Lett. **73**, 2090 (1998)
16. T. Yatsui, K. Itsumi, M. Kourogi, M. Ohtsu, Appl. Phys. Lett. **80**, 2257 (2002)
17. T. Yatsui, W. Nomura, M. Ohtsu, J. Nanophoton. **1**, 011550 (2007)
18. Y. Tanaka, K. Kobayashi, J. Micros. **229**, 228 (2008)
19. A. Sato, Y. Tanaka, F. Minami, K. Kobayashi, J. Lumines **129**, 1718 (2009)
20. N. Yasumaru, K. Miyazaki, J. Kiuchi, Appl. Phys. A **76**, 983 (2003)
21. S.Y. Chou, P.R. Krauss, P.J. Renstrom, Science **272**, 85 (1996)
22. A.N. Broers, J.M. Harper, W.W. Molzen, Appl. Phys. Lett. **33**, 392 (1978)
23. T. Ito, S. Okazaki, Nature **406**, 1027 (2000)

Chapter 2
Controlling the Size and Position in Nanoscale

2.1 Introduction

To realize the nanometer-scale controllability in size and position, we demonstrate the feasibility of nanometer-scale chemical vapor deposition using optical near-field techniques. These results are the basis for probe-less fabrication process described in the next section.

2.2 Fabrication of Nano-structure Using Optical Near-Field

2.2.1 Photo Chemical Vapor Deposition

In prior to the near-field optical chemical vapor deposition, here is the introduction of photo chemical vapor deposition (PCVD). PCVD is a promising method that can lower the growth temperature by forming the reactive radicals gallium and nitrogen via precursor photolysis. As an example of the deposition using PCVD, GaN has attracted much attention because of its excellent optical and electrical properties for wide-gap optical device and high-power and high-frequency electronic device applications [1–4]. For these applications, GaN is grown on AlN or GaN buffer layers on sapphire substrates using metalorganic chemical vapor deposition (MOCVD) or molecular beam epitaxy (MBE) [5–8]. These processes require high growth temperatures to obtain high-quality crystallinity in GaN. GaN, however, requires low temperatures during growth to reduce the thermal damage in substrates, as well as after growth during the cooling process to prevent cracking and thermal residual strain in the GaN film. Therefore, many different techniques have been proposed, such as pulsed laser deposition (PLD), ultrahigh-vacuum sputtering, and PCVD [9–12]. However, the detailed optical properties of GaN grown at room-temperature (RT) by these techniques remain unclear.

T. Yatsui, *Nanophotonic Fabrication*, Nano-Optics and Nanophotonics,
DOI 10.1007/978-3-642-24172-7_2, © Springer-Verlag Berlin Heidelberg 2012

GaN samples were grown on a sapphire (0001) substrate at room temperature [13–15]. Trimethylgallium (TMG) and semiconductor grade (99.999%) NH_3 as the III and V sources, respectively. H_2 was used as the carrier gas for the TMG. The flow rates of TMG and NH_3 were $0.25\,cm^3$/min and $2,000\,cm^3$/min, respectively. Thus, the V/III ratio was 8000:1. The total pressure in the reaction chamber was 5.4 Torr, and the deposition time was 60 min. Because the gas phases of TMG and NH_3 have strong photoabsorption at $\lambda < 270$ nm and $\lambda < 220$ nm, respectively [16,17]) we used a fifth harmonic of Q-switched Nd:YAG laser light ($\lambda = 213$ nm) as the light source for the photodissociation of the precursors. The PL spectra of the samples were examined using a continuous wave He–Cd laser ($\lambda = 325$ nm) (Fig. 2.1).

The morphology of the GaN sample was investigated with a scanning electron microscope (SEM). Figure 2.2a shows the overall SEM image of the GaN film on the substrate. Two typical areas, labeled area A and area B, were observed. In area A (see Fig. 2.2b, c), nanocrystallines could be seen, and the grain size ranged from 50 nm to 200 nm. In contrast, in area B (see Fig. 2.2d, e) dendrite-like structures with 100 nm widths were observed.

To examine the crystallinity, the optical properties of the deposited film were determined. First, the temperature-dependent PL spectra in area A was investigated (see Fig. 2.3a). Figure 2.3b shows the PL spectrum at 5 K from Fig. 2.3a, where a high and sharp peak at 3.47 eV (I_2) was observed. According to the literature [18, 19], the PL peak at 3.47 eV was derived from the bound exciton in hexagonal gallium nitride (h-GaN). As the temperature increased, the PL intensity of the peak I_2 decreased drastically, and it almost disappeared at temperatures above 100 K. In contrast, a new peak (labeled I_{FX}) emerged at temperatures above 100 K, and it remained at RT. These behaviors presumably resulted from the recombination of the neutral donor bound exciton to the free exciton in h-GaN [20]. Peaks I_2 and I_{FX} shifted with temperature, following Varshni's equation (dashed line in Fig. 2.3d), which defines the temperature dependence of the band gap [20, 21]. These results indicated that the GaN film synthesized by PCVD at RT had the same high quality of crystallinity as a GaN film synthesized using MOCVD at a growth temperature of 1,000°C and with an AlN buffer layer [22].

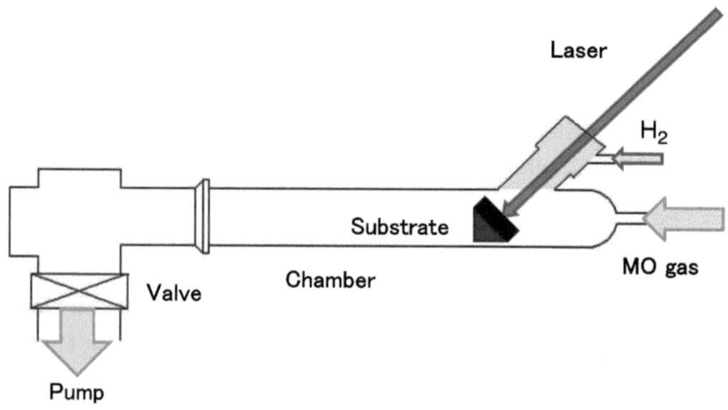

Fig. 2.1 Schematic of PCVD chamber

Fig. 2.2 SEM images of (**a**) the irradiated area. (**b**), (**c**) Typical surface morphology in area A. (**d**), (**e**) Typical surface morphology in area B

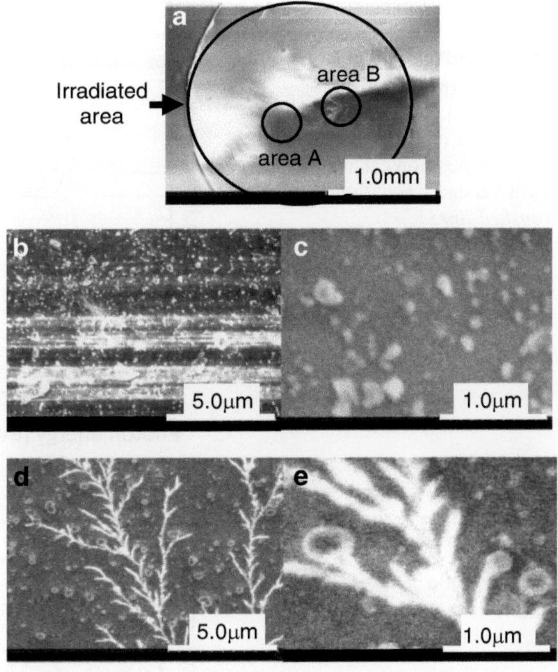

We evaluated the thermal strain in the GaN film using the value of the PL peak energy at RT. Zhao et al. proposed that the band gap of GaN at RT is expressed in terms of strain according to

$$E_g = 3.4282 + 0.0211\sigma \tag{2.1}$$

where E_g is the PL peak energy of a free exciton at RT and σ is the residual stress in the a-axis direction in GaN films, respectively [23]. Figure 2.3c shows the PL spectrum in Fig. 2.2a at 300 K in area A. E_g was 3.40 eV. With this value, the residual stress of the a-axis direction in GaN sample was 1.19 GPa. The values of residual stresses of GaN films with various buffer layers are listed in Table 2.1 [23–25]. Without a buffer layer, the residual stress was estimated to be 3.15 GPa due to the mismatch in the thermal expansion coefficients between GaN and the sapphire substrate [24]. The value of the residual stress in our sample was as low as that of a GaN film on a sapphire substrate that had a GaN buffer layer grown by MOCVD. These results confirmed that the growth of GaN film using PCVD at RT effectively relieved the thermal strain in GaN/sapphire due to the RT growth.

Second, we obtained temperature-dependent PL spectra from area B of dendrite-like GaN (Fig. 2.4a). In addition to the presence of I_2 at 3.47 eV, a blueshifted, sharp PL peak (I_{QW} line) at 3.55 eV was observed at 5 K, indicating the quantum size effect in dendrite-like GaN. Although the SEM photo did not help to determine the

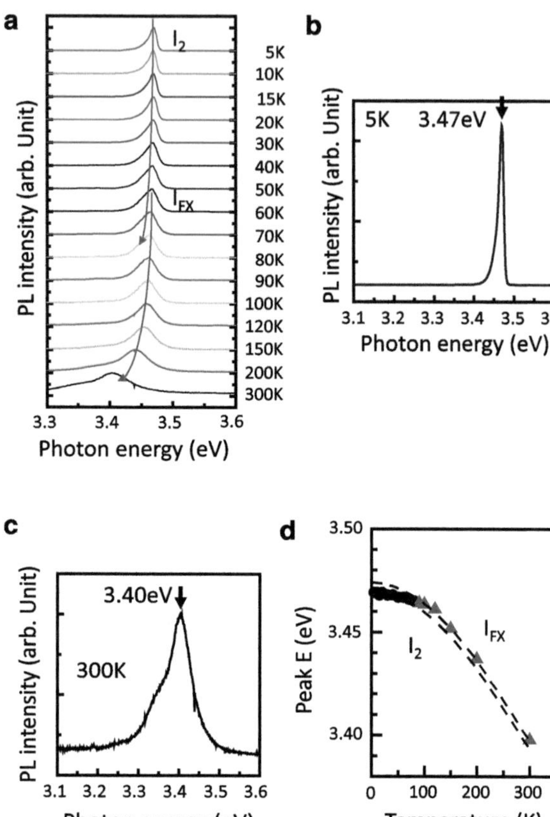

Fig. 2.3 (**a**) Temperature-dependent PL spectra from area A, and the same PL spectra in (**a**) obtained at (**b**) 5 K and (**c**) 300 K. (**d**) Temperature-dependent PL peak position calculated with Varshni's equation (*dashed line*)

Table 2.1 The residual stress values of GaN films

Bufferlayer	Substrate	Residual stress [GPa]	Reference
Withoutbufferlayer	3.15	Sapphire	[24]
GaN	Sapphire	1.16	[24]
AlN	Si	1.1	[23]
PorousGaN	GaAs/Si	1.4	[25]
Thiswork	Sapphire	1.19	

thickness of dendrite-like GaN, the value of the spectral shift from I_2 corresponds to a thickness of 3 nm. Figure 2.4b shows the PL peak position as a function of temperature. At low temperatures ($T < 60$ K), the spectra were dominated by neutral donor bound exciton recombination. As the temperature increased, I_{QW} progressively decreased and therefore benefited free exciton recombination (I_{FX} line). The PL peaks of I_{QW} and I_{FX} did not follow Varshni's equation (dashed line in Fig. 2.4b) at

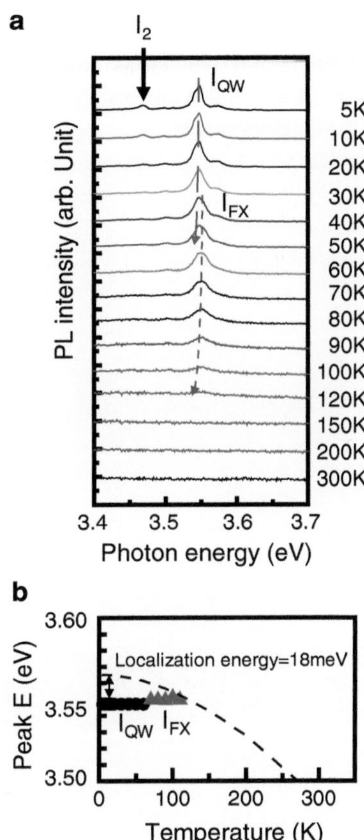

Fig. 2.4 (**a**) Temperature-dependent PL spectra and (**b**) PL peak position from area B. *Dashed line* shows the emission peak calculated with Varshni's equation

temperatures lower than 100 K. The energy difference between the estimated value using Varshni's equation and the observed value was as large as 18 meV, which corresponds to the excitonic localization energy reported in a GaN quantum well [26, 27]. As the temperature exceeded 120 K, we saw no spectral peaks. Because our synthesized GaN did not have any cap layers, the free exciton was not stable, and the nonradiative recombination process dominated above 120 K. The formation of the dendrite-like GaN nanostructure was attributable to its growth at RT using PCVD, when the low kinetic energy of the precursors for migration aided the growth.

Detailed observations in area B (Fig. 2.2) showed that these dendritic GaN nanostructures were similar in shape to those that arose from the two-dimensional diffusion-limited aggregation (DLA) model (Fig. 2.5a–c). DLA model was introduced by Witten and Sander [28, 29]. The model possesses important features of the dendritic fractal structure growth process. The random dendritic patterns under the DLA model are derived from the solution of the Laplace equation with moving boundaries [30]. The relevance of using DLA to describe the formation of fractal objects in nature [31] has been demonstrated by many experiments in

Fig. 2.5 Typical SEM
images for dendritic GaN
nanostructures in area B in
Fig. 2.2 observed at different
(**a–c**) magnifications

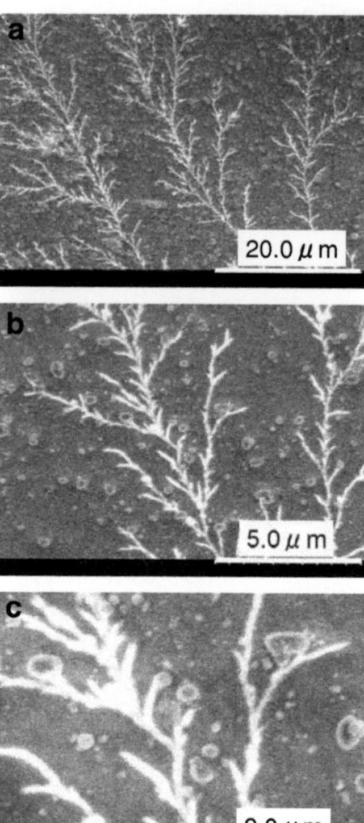

hydrodynamics [32–35], electrodeposition [36–39], and sputter deposition [40]. Thus, special attention has been paid to obtain dendrites, and various methods have been developed for fabricating dendritic nanostructures [41].

To compare the observed structures with those from the DLA model, we measured the fractal dimensions of the dendritic GaN nanostructures by binarizing the SEM images from Fig. 2.5a–c. The fractal dimensions D were calculated by the box-counting method. Figure 2.6a–c shows the results calculated from Fig. 2.5a–c and the relationship between the counted number n (L) of bright pixels and pixel size L. We estimated that $D = 1.59$ for the main branch in Fig. 2.5a, b and $D = 1.66$ for the main branch in Fig. 2.5c by using the degree of leaning from Fig. 2.6a–c, respectively. Although these SEM images had different magnifications, the fractal dimensions were comparable. These results indicate that the deposited GaN nanostructures had dendritic fractal structures. Furthermore, they were comparable to the fractal dimension from the DLA model with $D = 1.7$. The profiles of those SEM images were also similar to the DLA model simulation [41,42]. The similarity of the profiles was important for the fractal structure. This agreement between the fractal dimensions and appearances indicates that the dendritic GaN nanostructures grew by the DLA model growth process.

Fig. 2.6 The relationship between the counted number $n(L)$ of bright pixels and the pixel size L determined from the binarization SEM images from Fig. 2.5a–c and calculated by the box-counting method. From the calculated results the leans yielded the fractal dimensions of the main branches. (a), (b) $D = 1.59$ for the main branch in Fig. 2.5a, c, respectively. (c) $D = 1.66$ for the main branch in Fig. 2.5c

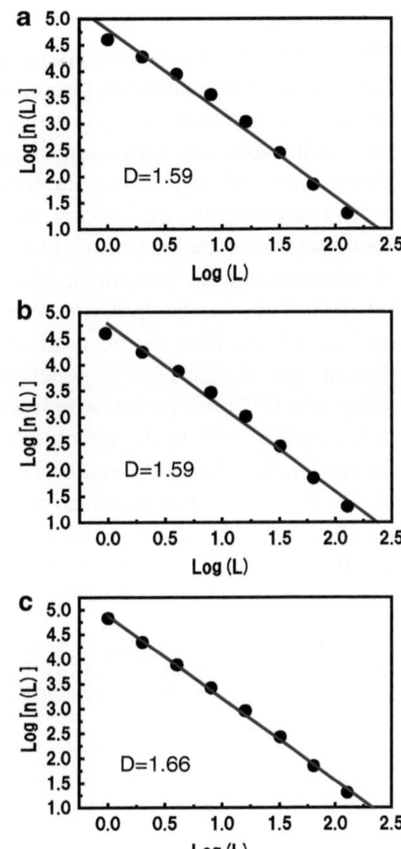

With the DLA growth model, the structure grew by the following three processes: (1) One particle approached from infinity by random walk; (2) If the particle reached the solid, the particle solidified; and (3) Repeat steps (1) and (2).

In the growth process a more acute branch tip angle promotes more effective solidification of the diffusing particles in the deposited structures and a higher growth rate. This is analogous to the motion of the solidification front in cooler gases. If the interface bulges into the gas or contacts the substrate, then the solidification front moves into a relatively cooler region and advances its growth. The DLA growth mechanism could be applicable to our experimental results. In our case, the formation of dendritic GaN nanostructures was attributed to RT growth using PCVD, in which the low kinetic energy of the migration precursors resulted in the growth. The kinetic energy was supplied while the rest of the photodissociation energy was irradiated by the light source. Because conventional CVD growth at temperatures exceeding 1,000°C results in uniform film growth, dendritic fractal structures originate from RT growth.

2.2.2 *Near-Field Optical Chemical Vapor Deposition*

For realization of nanoscale photonic device required by the future system, electron beams [43], and scanning probe microscopes [44, 45] have been used to control the site on the substrate. However, these techniques have a fatal disadvantage because they cannot deal with insulators, limiting their application.

To overcome this difficulties, in this section, we demonstrated near-field optical chemical vapor deposition (NFO-CVD, Fig. 2.7), which enables the fabrication of nanometer-scale structures, while precisely controlling their size and position [46–51]. That is, the position can be controlled accurately by controlling the position of the fiber probe used to generate the optical near field. To guarantee that an optical near field is generated with sufficiently high efficiency, we used a sharpened UV fiber probe, which was fabricated using a pulling/etching technique [52] (see Fig. 2.8). In the uncoated condition, the diameter of the sharpened probe tip remained sufficiently small. This enabled high-resolution position control and in-situ shear-force topographic imaging of the deposited nanometer-scale structures (see Fig. 2.9). Since the deposition time was sufficiently short, the deposition of metal on the fiber probe and the resultant decrease in the throughput of optical

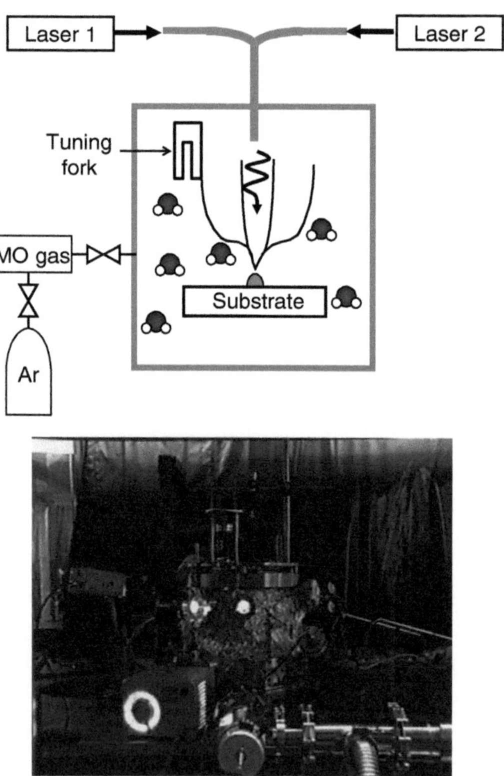

Fig. 2.7 Schematic and photograph of NFO-CVD

Fig. 2.8 Fabrication process
of fiber probe. (**a**) Pulling the
fiber. (**b**) Etching the fiber
using HF solution. (**c**) SEM
image of the sharpened
fiber probe

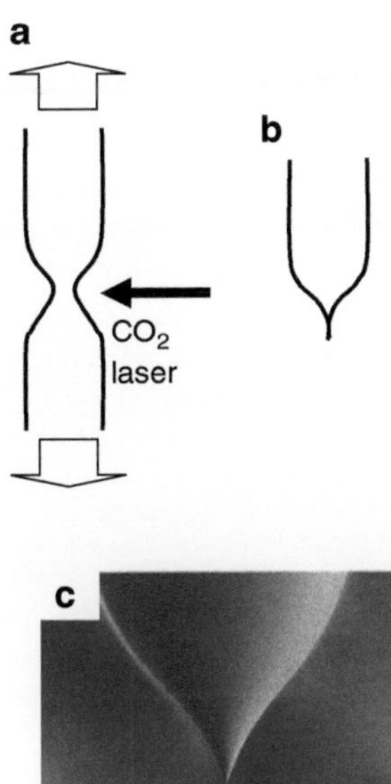

near-field generation were negligible. The separation between the fiber probe
and the sapphire (0001) substrate was kept within a few nanometers by shear-
force feedback control. Immediately after the nano-dots were deposited, their
sizes and shapes were measured by in-situ vacuum shear-force microscopy [46],
using the same probe as used for deposition. Due to the photochemical reaction
between the reactant molecules and the optical near field generated at the tip of an
optical fiber probe, NFO-CVD is applicable to various materials, including metals,
semiconductors, and insulators.

Figure 2.10 shows the schematic of the shear-force feedback control system
[53, 54]. The typical resonant frequency of the quartz tuning fork was 32.768 Hz
(Γ_q), the internal resistance was 35 kΩ, and the spring constant of the quartz tuning
fork was 23.5 kN/m [53]. To realize higher resonant frequency of the fiber than Γ_q,
the required length of the fiber probe from the edge of the quartz tuning fork, l is

$$l < \lambda \sqrt{\left(\frac{d}{8\pi}\right)\sqrt{\frac{E}{\rho}}} \sim 500\,(\mu m), \qquad (2.2)$$

where λ is the 1st resonant mode $= 1.875$, d the fiber cladding diameter $= 125\,\mu$m,
and ρ is the density $= 2{,}650\,$kg/m^3 [55]. Based on the calculation, the fiber length
was set as 150 μm (see the inset of Fig. 2.10). Figure 2.11 shows the frequency
spectra of the quartz tuning fork with fiber probe, and the resonant frequency of

Fig. 2.9 Schematic and
photograph of shear-force
microscopy

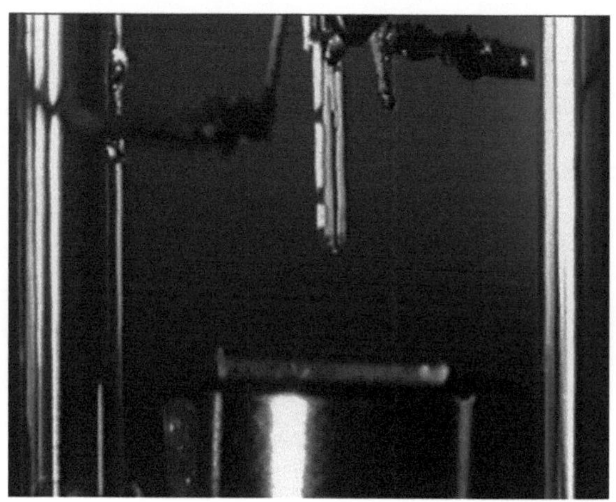

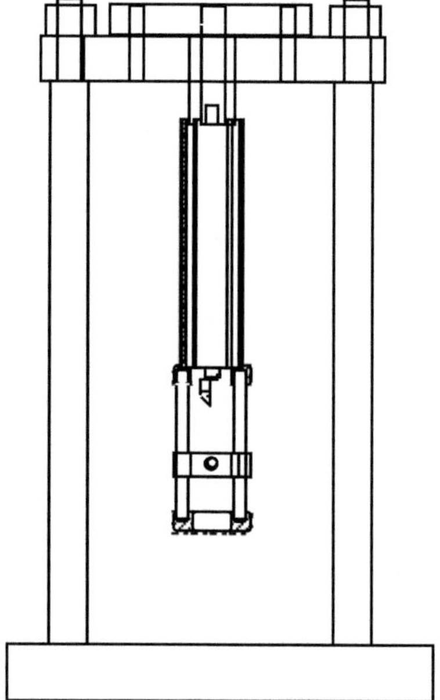

32.430 Hz with Q value of 1,500, where the sensitivity (S) [55] and mechanical
time constant (τ) were obtained as follows:

$$S = \frac{V_{\text{out}}}{x} = \frac{24\pi f R d k L_{\text{e}} (L_{\text{e}}/2 - L)}{t^2} \sim 600 \, (\mu\text{V/nm}), \qquad (2.3)$$

$$\tau = \frac{2Q}{2\pi f} \sim 20 \, (\text{ms}), \qquad (2.4)$$

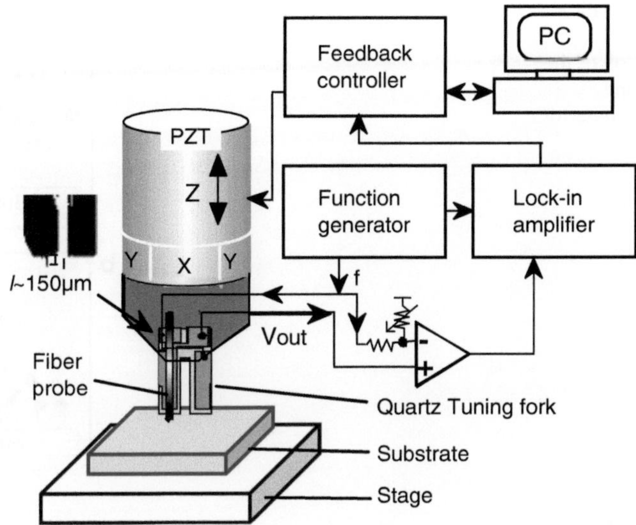

Fig. 2.10 Schematic of the shear-force feedback control system

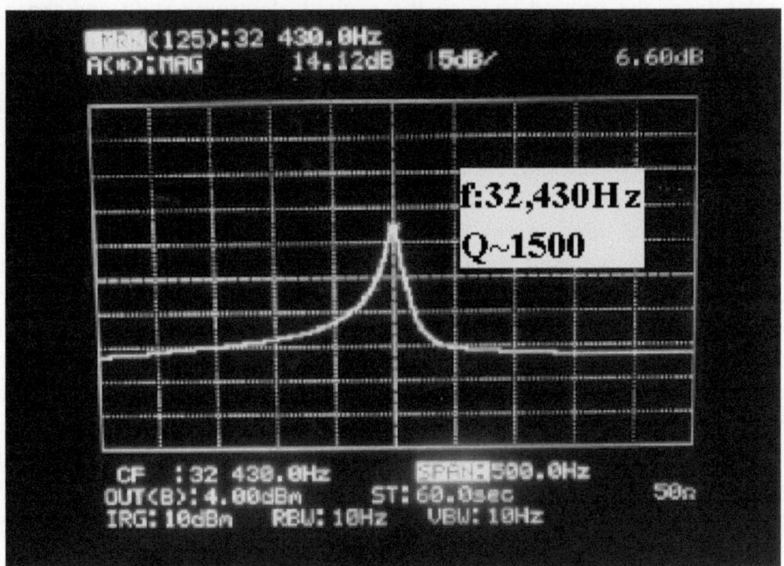

Fig. 2.11 Schematic of the shear-force feedback control system

where x is oscillation amplitude and L_e is the length of the electrode of the quartz tuning fork.

Conventional optical CVD method uses a light source that resonates the absorption band of megalomaniac (MO) vapor and has a photon energy that exceeds

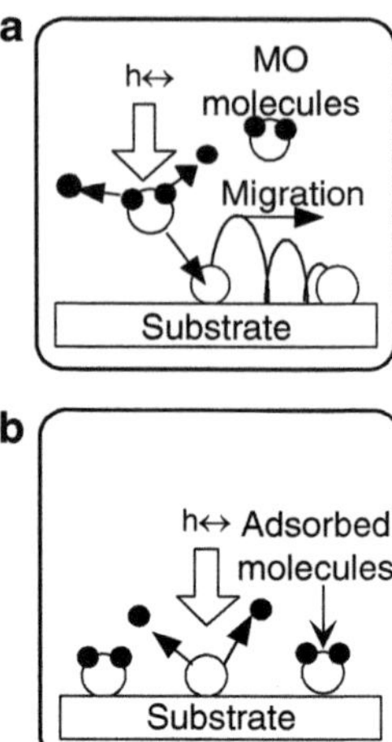

Fig. 2.12 Schematic diagrams of the photodissociation of the (**a**) gas-phase and (**b**) adsorption-phase MO molecules

the dissociation energy [58]. Thus, it utilizes a two-step process; gas-phase photodissociation and subsequent adsorption. In this process, resonant photons excite molecules from the ground state to the excited electronic state and the excited molecules relax to the dissociation channel, and then the dissociated metallic atoms adsorb to the substrate [59]. However, we found that the dissociated MO molecules migrate on the substrate before adsorption, which limits the minimum lateral size of deposited dots (Fig. 2.12a). A promising method for avoiding this migration is dissociation and deposition in the adsorption-phase (Fig. 2.12b).

As an example of NFO-CVD, we introduce the deposition of Zn dot. Since the absorption band edge energy of the gas-phase diethylzinc (DEZn) was 4.6 eV ($\lambda = 270$ nm) [58], we used He–Cd laser light (3.81 eV, $\lambda = 325$ nm) as the light source (laser 1 in Fig. 2.7) for the deposition of Zn; it is nonresonant to gas-phase DEZn. However, due to the red-shift in the absorption spectrum in DEZn of adsorption-phase with respect to that in the gas-phase, i.e., it resonates the adsorption-phase DEZn. The red-shift may be attributed to perturbations of the free-molecule potential surface in the adsorbed-phase [58, 60]. Using a sharpened UV fiber probe, we achieved selective dissociation of adsorbed DEZn, as a results, we successfully fabricated 20-nm Zn dots with 65-nm separation on sapphire (0001) substrate (see Fig. 2.13) [48, 50]. Furthermore, since the nonresonant propagating light that leaked from the probe did not dissociate the gas-phase DEZn, the atomic-level sapphire steps around the deposited dots were clearly observed after the deposition. By changing the reactant molecules, nanometric Al dots were deposited (see Fig. 2.14). In addition, Zn and Al dots were successively deposited on the same sapphire substrate with high precision (see Fig. 2.15) [51]. The deposited dots confirmed their composition by energy dispersive X-ray (EDX) spectra (see Fig. 2.16), in which in addition to the spectral peak of Si substrate, additional spectral peaks corresponding to the Zn and Al were confirmed.

Fig. 2.13 (**a**) Shear-force image of closely spaced Zn dots. (**b**) Cross-sectional profile

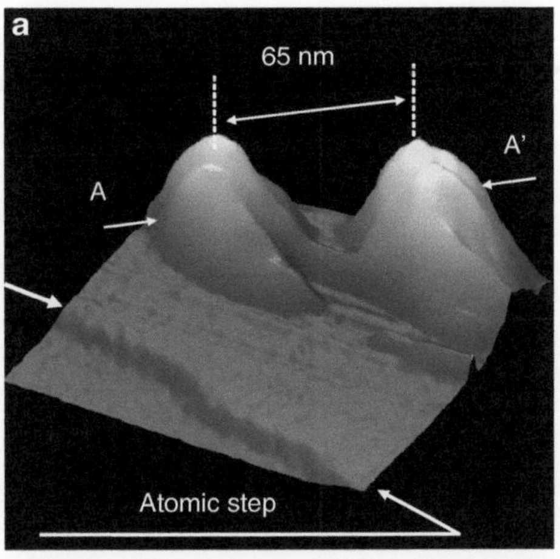

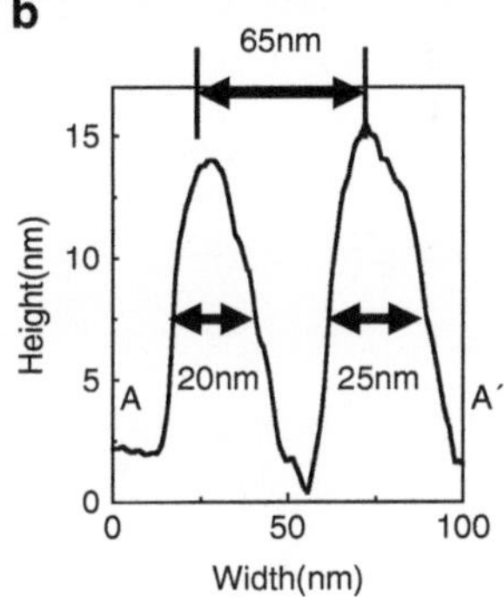

Since high-quality ZnO nanocrystallites can be obtained by oxidizing Zn nuclei [56, 57], NFO-CVD could be used to produce high-quality ZnO nanocrystallites; a promising material for use in nanometer-scale light-emitters and switching devices in nanophotonic integrated circuits. Furthermore, to confirm that the deposited dots were Zn, we fabricated a UV-emitting ZnO dot by oxidizing the Zn dot immediately after deposition [48].

First, the Zn dot was deposited using selective dissociation of adsorbed DEZn (Fig. 2.17a). Next, laser annealing was employed for this oxidization [57], i.e., the deposited Zn dot was irradiated with a pulse of an ArF excimer-laser ($\lambda = 193$ nm, pulse width: 30 ns, fluence: 120 mJ/cm^2) in a high-pressure oxygen environment (5 Torr). Finally, to evaluate the optical properties of the oxidized dot, the photo-luminescence (PL) intensity distribution was measured using an illumination and collection mode (IC-mode) near-field optical microscope. For this measurement, a He–Cd laser ($\lambda = 325$ nm) was used as the light source and the signal collected through a fiber probe and a long wave ($\lambda > 360$ nm) pass filter was focused on a photomultiplier tube (PMT) to count photons.

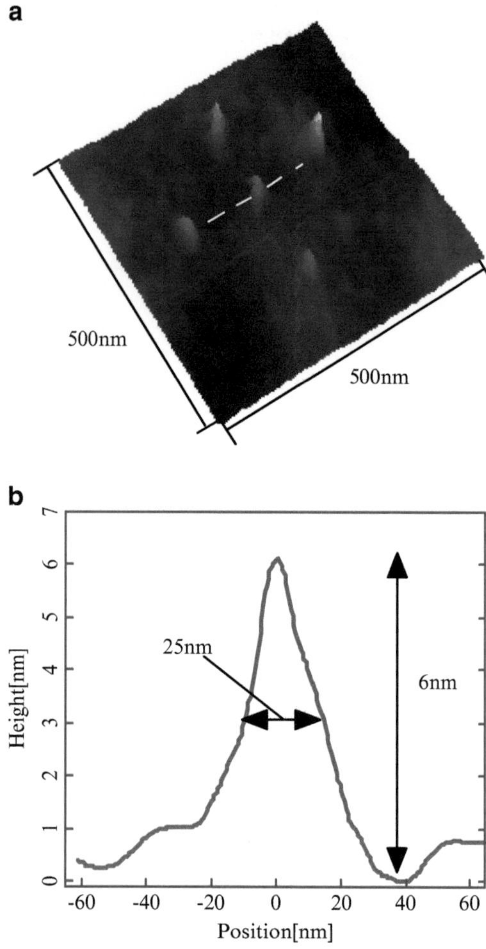

Fig. 2.14 (**a**) Shear-force image of closely spaced Al dots. (**b**) Cross-sectional profile

Figure 2.17b shows the PL intensity distribution ($\lambda > 360$ nm) of an oxidized dot. The low collection efficiency due to the IC-mode configuration did not establish the spectrum. Thus, a Zn thin film was deposited using the same CVD process, except the optical near field was replaced by far-field propagating light. After it was annealed with an excimer laser, we found that the PL intensity of the spontaneous emission from the free exciton was ten times greater than that of the deep-level green emission. Thus, we concluded that the PL in Fig. 2.17a originated from spontaneous emission from the free exciton. Figure 2.17c is the cross-sectional profile through the spot in Fig. 2.17b. Note that the full width at half maximum (FWHM) in Fig. 2.17c is smaller than the dot size (FWHM of 100 nm, see Fig. 2.17a), which was estimated using a shear-force microscope. This originates from the high spatial resolution capability of the IC-mode near-field microscope. The next stage of this study will be a more detailed evaluation of the optical properties of single ZnO nanocrystallites.

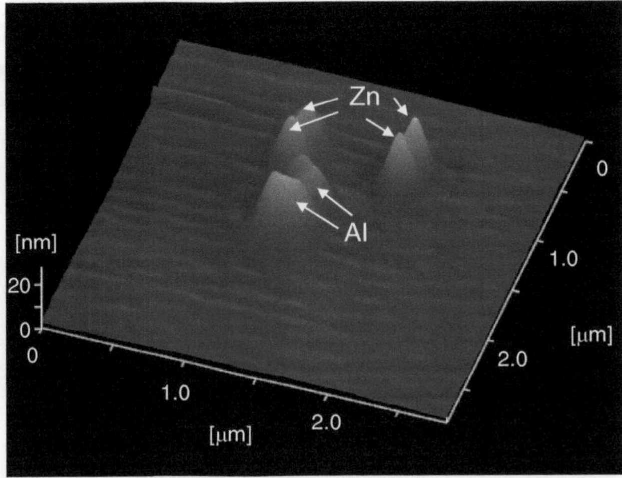

Fig. 2.15 Shear-force image of closely spaced (**a**) Zn dots and (**b**) Zn and Al dots

Fig. 2.16 EDX spectra at the
are of Zn (curve A) and Al
(curve B) deposited

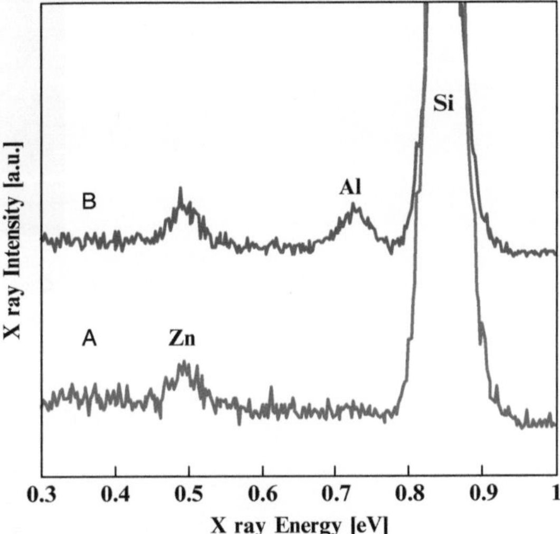

2.2.3   *Regulating the Size and Position of Deposited Nanoparticles*

In order to realize further controllability of size, in this section, we utilize the dependence of plasmon resonance on the photon energy of optical near fields and control the growth of Zn nanoparticles during the process of Zn deposition. Using this dependence, we demonstrate the deposition of a nanometer-scale dot using NFO-CVD [49].

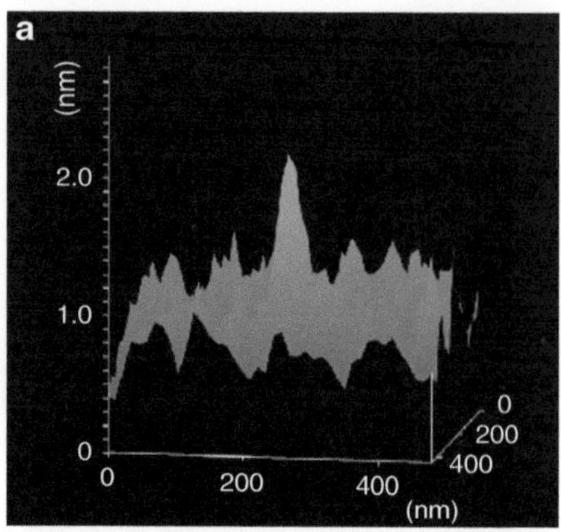

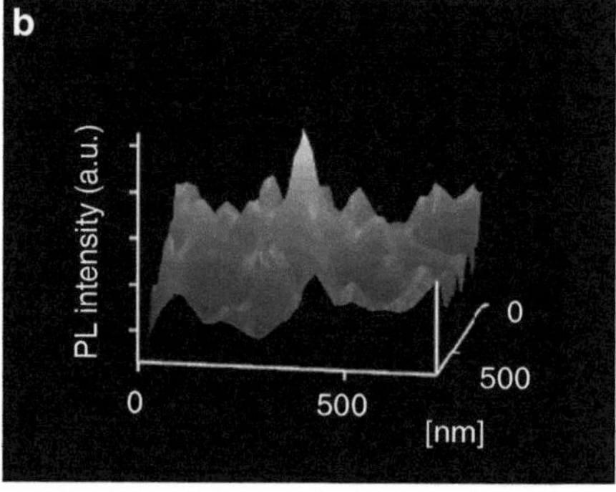

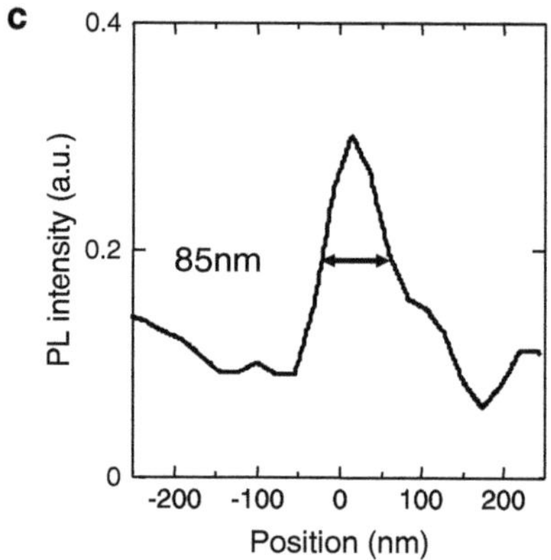

Fig. 2.17 (**a**) Shear-force image of Zn dot. (**b**) Spatial distribution of the optical near-field PL intensity ($\lambda >$ 360 nm) of an oxidized Zn dot [(**a**)]. The image is of a 750×750-nm area. (**c**) Cross-sectional profile through the spot of (**a**)

First, we studied nanoparticle formation on the cleaved facets of UV fibers (core diameter $= 10\ \mu$m) using conventional optical CVD (see Fig. 2.18a, b).

Gas-phase DEZn at a partial pressure of 5 mTorr was used as the source gas. The total pressure, including that of the Ar buffer gas, was 3 Torr. As the light source for the photodissociation of DEZn, a 500-μW He–Cd laser [photon energy

Fig. 2.18 Schematics of (**a**)
and (**b**) conventional optical
CVD on the cleaved facet of
an optical fiber and
(**c**) near-field optical CVD

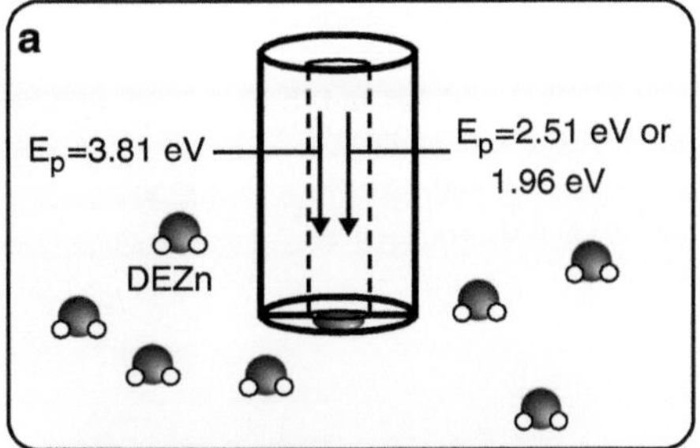

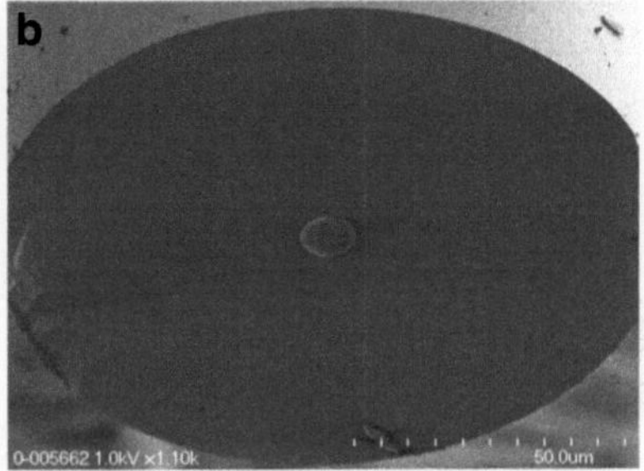

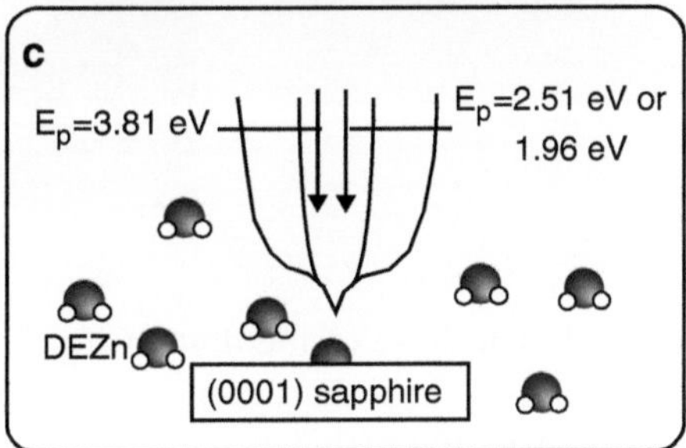

$E_p = 3.81\,\text{eV}$ ($\lambda = 325\,\text{nm}$)] was coupled to the other end of the fiber. The irradiation time was 20 s. This irradiation covered the facet of the fiber core with a layer of Zn nanodots. Figure 2.19 shows a scanning electron cartographic (SEM) image of the deposited Zn nanodots, and their size distribution is shown in Fig. 2.20. The peak diameter and FWHM of this curve are 110 nm and 50 nm, respectively.

In order to control the size distribution, we introduced 20 μW Ar$^+$ ($E_p = 2.54\,\text{eV}$ [$\lambda = 488\,\text{nm}$]) or He–Ne ($E_p = 1.96\,\text{eV}$ [$\lambda = 633\,\text{nm}$]) lasers into the fiber, in addition to the He–Cd laser. Their photon energies are lower than the absorption band edge energy of DEZn, i.e., they are nonresonant light sources for the dissociation of DEZn. The irradiation time was 20 s. Figure 2.19b, c shows SEM images of the Zn nanodots deposited with irradiation at $E_p = 3.81\,\text{eV}$ and 2.54 eV and at $E_p = 3.81\,\text{eV}$ and 1.96 eV, respectively. Figure 2.19e, f shows the respective size distributions. The peak diameters are 30 nm and 18 nm, respectively, which are

Fig. 2.19 SEM images of Zn nanoparticles deposited by optical CVD with (**a**) and (**b**) $E_p = 3.81$ eV, (**c** and **d**) $E_p = 3.81$ eV and 2.54 eV, and (**e**) and (**f**) $E_p = 3.81$ eV and 1.96 eV

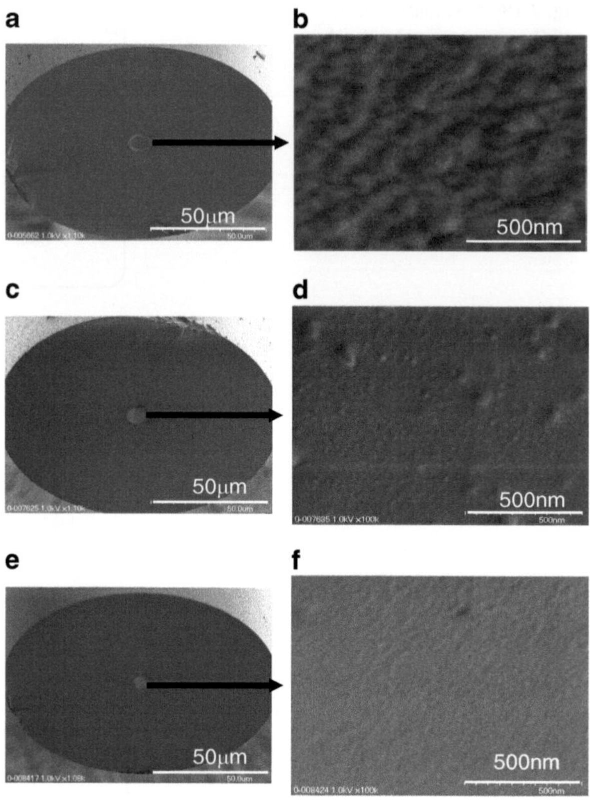

smaller than those of the dots in Fig. 2.19d, and depend on the photon energy of the additional light. Furthermore, the FWHM (10 nm and 12 nm, respectively) was definitely narrower than that of Fig. 2.19d. These results suggest that the additional light controls the size of the dots and reduces the size fluctuation, i.e., size regulation is realized (Fig. 2.21).

We now discuss the possible mechanisms by which the additional light regulates the size of the dots. A metal nanoparticle has strong optical absorption due to plasmon resonance [32, 62], which strongly depends on particle size. This can induce the desorption of the deposited metal nanoparticles [33, 34]. As the deposition of metal nanoparticles proceeds in the presence of light, the growth of the particles is affected by a trade-off between deposition and desorption, which determines their size, and depends on the photon energy. It has been reported that surface plasmon resonance in a metal nanoparticle is red-shifted with increasing the particle size [33, 34]. However, our experimental results disagree with these reports (compare Fig. 2.19d–f). In order to find the origin of this disagreement, a series of calculations were performed and resonant sizes were evaluated. Mie's theory of scattering by a Zn sphere was employed, and only the first mode was considered [65]:

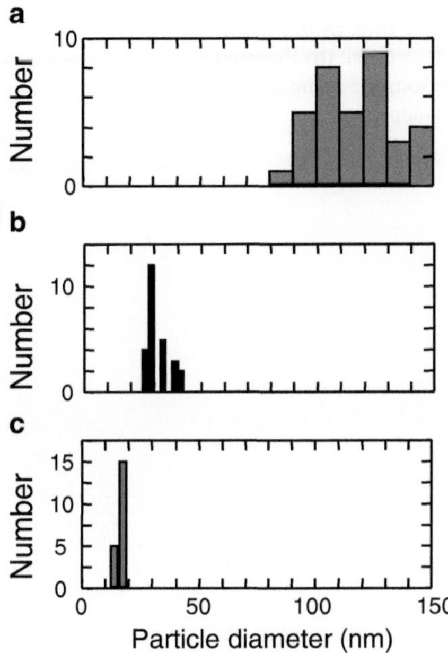

Fig. 2.20 The diameter distributions of the Zn particles; (**a**), (**b**), and (**c**) are for (**b**), (**d**), and (**f**) of Fig. 2.19, respectively

$$\alpha = \frac{1 - \frac{1}{10}\left(\varepsilon + \varepsilon_m\right)x^2 + O\left(x^4\right)}{\left(\frac{1}{3} + \frac{\varepsilon_m}{\varepsilon - \varepsilon_m}\right) - \frac{1}{30}\left(\varepsilon + 10\varepsilon_m\right)x^2 - i\frac{4\pi^2\varepsilon_m{}^{3/2}}{3}\frac{V}{\lambda_0{}^3} + O\left(x^4\right)}V, \qquad (2.5)$$

where $x\left(= \pi a_{\text{sphere}}/\lambda_0\right)$ is the size parameter, with a_{sphere} the diameter and λ_0 the wavelength in vacuum. The curves in Fig. 2.22a represent the calculated polarizability α with respect to three photon energies. The vertical axis is the value of α normalized to the volume, V, of a Zn sphere in the air, which depends on its diameter and is maximal at a certain diameter, i.e., at the resonant diameter. The solid curve in Fig. 2.22b represents the resonant diameter as a function of the photon energy, which is not a monotonous function and takes the minimum at $E_p = 2.0\,\text{eV}$ ($\lambda = 620\,\text{nm}$). Since the imaginary part of the refractive index of the Zn also takes a maximum also at $E_p = 2.0\,\text{eV}$ ($\lambda = 620\,\text{nm}$) (see the broken curve in Fig. 2.22b) [66], the minimum of the solid curve is due to the strong absorption in Zn.

Although Fig. 2.22a shows that the resonant diameter (95 nm) for $E_p = 2.54\,\text{eV}$ exceeds that (80 nm) for $E_p = 3.81\,\text{eV}$, the calculated resonant diameter for $E_p = 3.81\,\text{eV}$ is in good agreement with the experimentally confirmed particle size (see curve A in Fig. 2.19d). Since the He–Cd laser light ($E_p = 3.81\,\text{eV}$) is resonant for the dissociation of DEZn and is responsible for the deposition, irradiation with a He–Cd laser during deposition causes the particles to grow, and this growth halts when the particles reach the resonant diameter, because the rate of desorption increases due to resonant plasmon excitation. This is further supported by the fact that the resonant diameter (75 nm) for $E_p = 1.96\,\text{eV}$ is smaller than that for

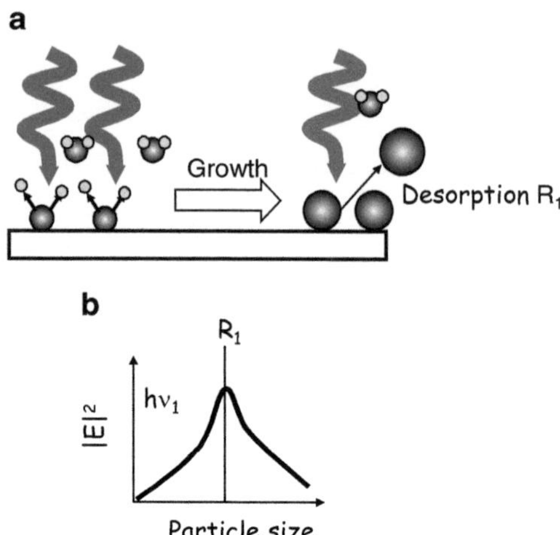

Fig. 2.21 (**a**) The size regulation by the near-field desorption. (**b**) Plasmon resonance in the metal nanoparticle

$E_\mathrm{p} = 3.81\,\mathrm{eV}$ (see Fig. 2.22a) and illumination with the additional light causes the particles to shrink (see Fig. 2.19d).

Another possible mechanism involves the acceleration of dissociation by the additional light. The photodissociation of DEZn produces transient monoethylzinc; then, Zn results from the dissociation of monoethylzinc. Although the absorption band of monoethylzinc was not determined, the photon-energy dependence of the size observed using the additional light might be due to the acceleration of the dissociation rate, i.e., the additional light, which is nonresonant for DEZn, resonates the monoethylzinc [67], since the first metal-alkyl bond dissociation has a larger dissociation energy than the subsequent metal-alkyl bond dissociation [68, 69].

Finally, using this dependence, we used NFO-CVD (see Fig. 2.7) to control the position of the deposited particle. Figure 2.23a–c shows topographical images of Zn deposited by NFO-CVD with illumination with a $1\text{-}\mu\mathrm{W}$ He–Cd laser ($E_\mathrm{p} = 3.81\,\mathrm{eV}$, laser 1 in Fig. 2.7) alone, or together with a $1\text{-}\mu\mathrm{W}$ Ar$^+$ laser ($E_\mathrm{p} = 2.54\,\mathrm{eV}$) or a $1\text{-}\mu\mathrm{W}$ He–Ne laser ($E_\mathrm{p} = 1.96\,\mathrm{eV}$) (laser 2 in Fig. 2.7), respectively. The irradiation times were 60 s. During deposition, the partial pressure of DEZn and the total pressure including the Ar buffer gas were maintained to 100 mTorr and 3 Torr, respectively. In Fig. 2.23d, curves A, B, and C are the respective cross-sectional profiles through the Zn dots in Fig. 2.23a–c. The respective FWHM was 60 nm, 30 nm, and 15 nm; i.e., a lower photon energy gave rise to smaller particles, which is consistent with the experimental results shown in Fig. 2.19.

These results suggest that the additional light controls the size of the dots and reduces the size fluctuation, i.e., size regulation is realized. Furthermore, the position can be controlled accurately by controlling the position of the fiber probe used to generate the optical near field. The experimental results and the suggested mechanisms described above show the potential advantages of this technique in improving the regulation of size and position of deposited nanodots. Furthermore,

Fig. 2.22 (**a**) Curves *A*, *B*, and *C* show the calculated polarizability α-normalized to the volume *V* for a Zn sphere surrounded by air for $E_p = 3.81$ eV, 2.51 eV, and 1.96 eV, respectively. (**b**) The resonant diameter of a Zn sphere (*solid curve*). The imaginary part of the refractive index of Zn, n_{Zn}, used for the calculation (*broken curve*) (see [66])

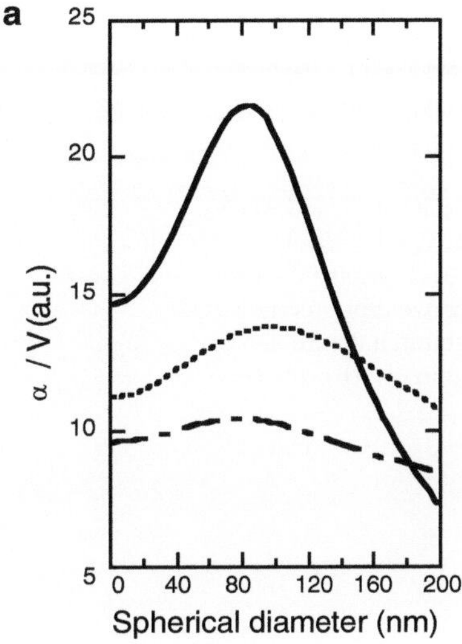

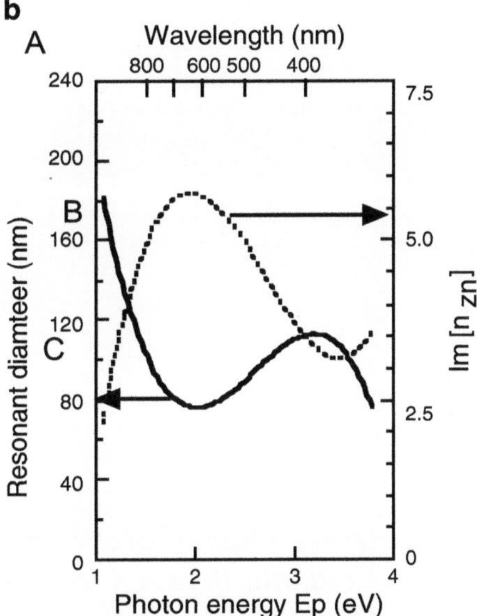

since our deposition method is based on a photodissociation reaction, it could be widely used for nanofabrication of the other material for example GaN, GaAs, and so on.

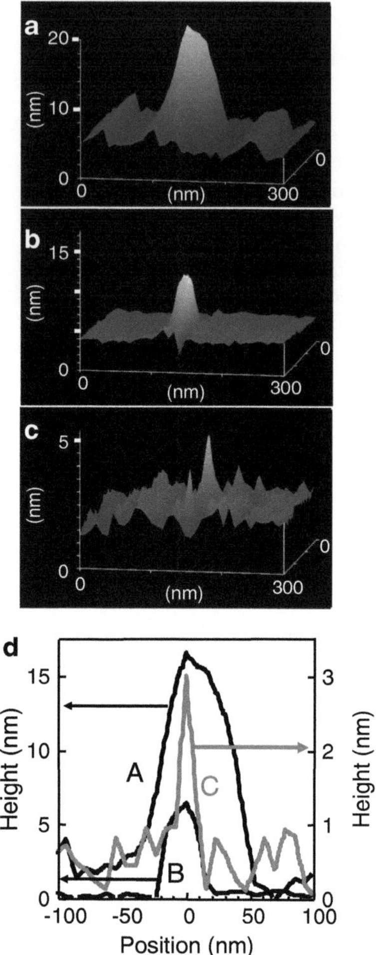

Fig. 2.23 Bird's-eye views of shear-force topographical images of Zn deposited by NFO-CVD with (**a**) $E_p = 3.81$ eV, (**b**) $E_p = 3.81$ eV and 2.54 eV, and (**c**) $E_p = 3.81$ eV and 1.96 eV, respectively. (**d**) Curves A, B, and C show the respective cross-sectional profiles through the Zn dots deposited in (**a**), (**b**), and (**c**)

2.2.4   Observation of Size-Dependent Resonance of Near-Field Coupling Between Deposited Zn Dot and Probe Apex During NFO-CVD

To realize sub-10-nm scale controllability in size, we report here the precise growth mechanism of Zn dots with NFO-CVD. We directly observe that the deposition rate is maximal when the dot grew to a size equivalent to the probe apex diameter. This dependence is well accounted for by the theoretically calculated dipole-dipole coupling with a Förster field. The theoretical support and experimental results indicate that potential advantages of this technique for improving regulation of the size and position of deposited nanometer-scale dots.

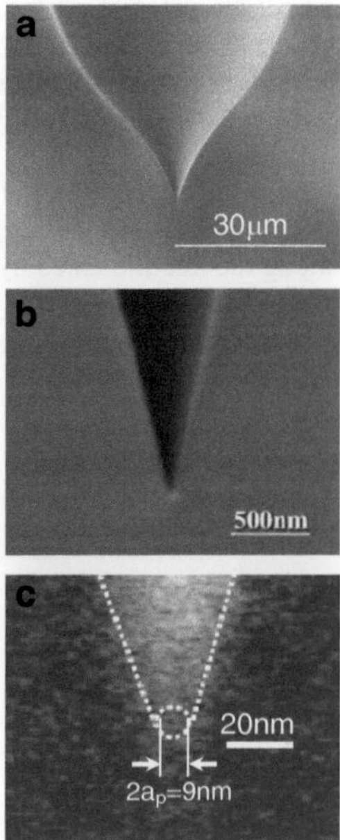

Fig. 2.24 A SEM image of
(**a**) UV fiber probe, (**b**)
magnified image of (**a**), and
(**c**) magnified image of (**c**).
$2a_p$: apex diameter

Figure 2.24b shows the SEM image of the fiber probe used in this study. The apex diameter $2a_p$ is estimated as 9 nm by referring to the fitted broken circle.

Figure 2.25a shows a shear-force image of four Zn dots deposited with the irradiation time of 60 s (dot A), 30 s (dot B), 10 s (dot C), and 5 s (dot D) with the laser output power P of 5 μW. We tried to deposit dots with 260 nm separations along the x axis by servo-controlling the position of the fiber probe. As shown in the cross-sectional profiles in Fig. 2.25b, Zn dots as small as 20 nm in their size S (defined by the FWHM on this profile) were fabricated. Their separation are 269 nm and 262 nm, by which high accuracy in position (< 10 nm) was confirmed. Major origin of this residual inaccuracy is the hysteresis of the PZT actuator used for scanning the fiber probe, which can be decreased by carefully selecting the actuator.

Figure 2.26 shows the normalized deposition rate R of Zn dots, as a function of the dot size S. Since the measured dot size S' was convolution of probe apex diameter $2a_p$ and the real size S, S was estimated as $S = S' - 2a_p$. It should be noted that R takes the maximum at $S = 2a_p$ (see Fig. 2.27). This result indicates that the magnitude of the near-field optical interaction between the deposited Zn dot and the probe apex is enhanced resonantly with respect to S, resulting in the resonant increase in R. In other words, the near-field optical interaction exhibits the size-dependent resonance characteristics.

To find the origin of this size-dependent resonance, we calculated the magnitude of the near-field optical interaction between closely spaced nanoparticles (Fig. 2.26b). The spheres p and s represent the probe apex and the Zn dot, respectively. Since the separation between two particles is much narrower than the wavelength, the Förster field (proportional to R^{-3}, R is the distance from the dipole)

Fig. 2.25 (**a**) A shear-force image of deposited Zn dots. The laser irradiation time of dots $A - D$ were 60 s, 30 s, 10 s, and 5 s, respectively. (**b**) Upper and lower curves show the cross-sectional profile along the line indicated by *arrows i–i'* and *ii–ii'*, respectively

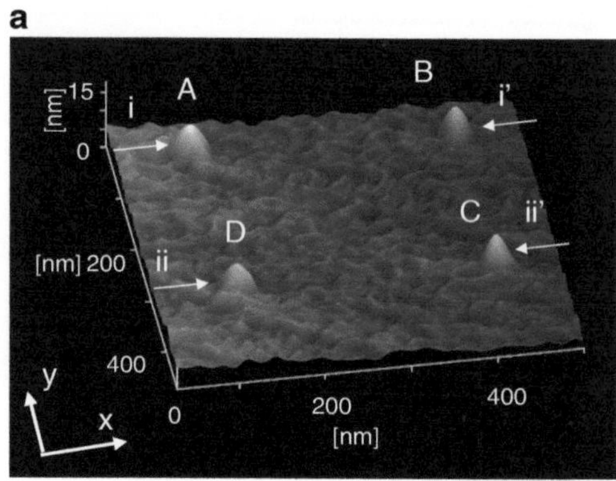

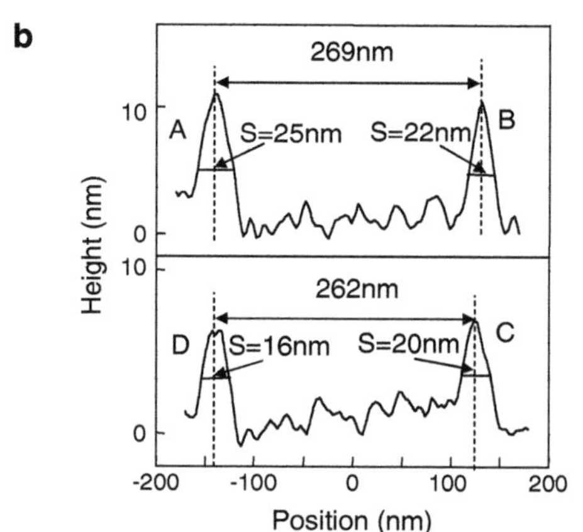

Fig. 2.26 The time dependent deposition rate R. *Solid squares and circles* indicate the normalized deposition rate p with $10\,\mu$W and $5\,\mu$W, respectively. *Solid curve* indicates the calculated value of I_2/I_1

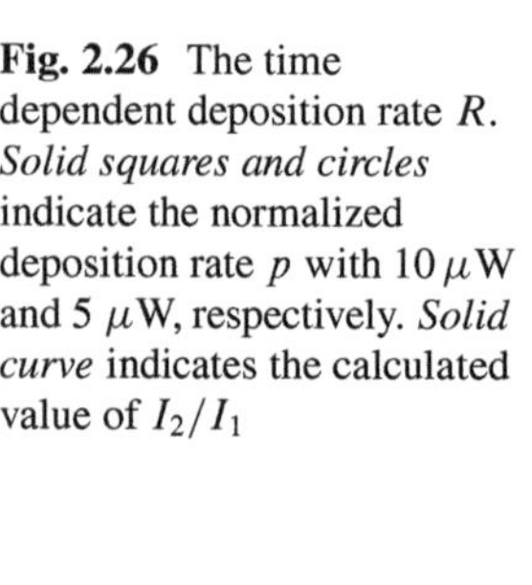

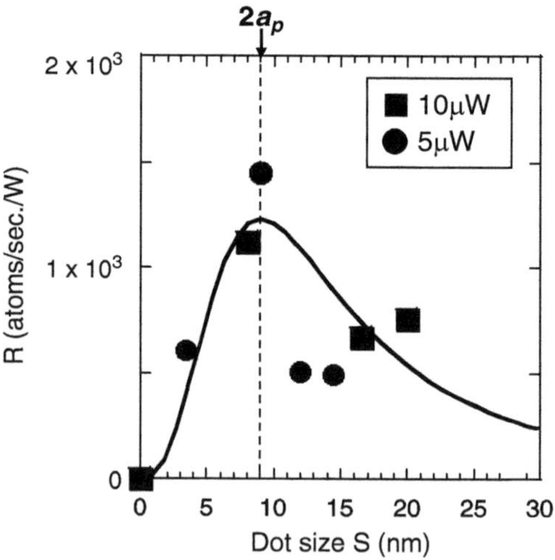

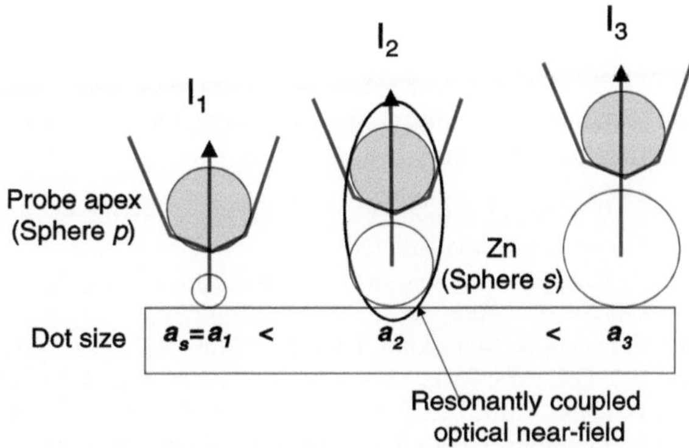

Fig. 2.27 Schematic of growth process of Zn dot

is dominant in the oscillating dipole electric field. In this quasistatic model, the intensity I_s of the light scattered from the two closely spaced spheres p and s is given by [70]

$$I_s = I_1 + I_2 = \left(\alpha_p + \alpha_s\right)^2 |E|^2 + 4\Delta\alpha \left(\alpha_p + \alpha_s\right) |E|^2 , \qquad (2.6)$$

where $\alpha_i = 4\pi\varepsilon_0(\varepsilon_i - \varepsilon_0)/(\varepsilon_i + 2\varepsilon_0)\, a_i^3$ is a polarizability of the sphere i $(=p, s)$ with diameter a_i. The first and second terms I_1 and I_2 represent the light intensity scattered from the spheres, and the light due to the dipole–dipole interaction induced by the Förster field. Thus, the light intensity under study, normalized to I_1, is given by

$$\frac{I_2}{I_1} = \frac{G_p A_p^3}{(A_p + 1)^3(G_p A_p^3 + 1)} \qquad (2.7)$$

where $A_p = a_p/a_s$ and $G_p = (\varepsilon_p - 1)(\varepsilon_s + 2)/(\varepsilon_p + 2)(\varepsilon_s - 1)$. For the deposition by the fiber probe, dielectric constant of Zn and fiber probe are $\varepsilon_s = (0.6 + i4)^2$ [71] and $\varepsilon_p = 1.5^2$, respectively. The diameter $2\,a_p$ of the sphere p was 9 nm (see Fig. 2.25b). Solid curve in Fig. 2.26a shows the calculated value of I_2/I_1 as a function of the Zn dot size $S(=2a_s)$, which agrees very well with the experimental results. This agreement indicates that the increase in R is originated from the dipole–dipole coupling with Förster field at the dot size equivalent to the probe apex diameter.

References

1. S. Nakamura, M. Senoch, S. Nagahama, N. Iwasa, T. Yamada, T. Matsushita, H. Kiyoku, Y. Sugimoto, Jpn. J. Appl. Phys. **35**, L217 (1996)
2. I. Akasaki, H. Amano, Jpn. J. Appl. Phys. **36**, 5393 (1997)
3. M.A. Khan, Q. Chen, M.S. Shur, B.T. Dermott, J.A. Higgins, J. Burm, W. Schaff, L.F. Eastman, Electron. Lett. **32**, 357 (1996)

4. Y.F. Wu, B.P. Keller, S. Keller, N.X. Nguyen, M. Le, C. Nguyen, T.J. Jenkins, L.T. Kehias, S.P. DenBaars, U.K. Mishra, IEEE Electron. Device Lett. **18**, 438 (1997)
5. H. Amano, M. Kito, K. Hiramatsu, I. Akasaki, Jpn. J. Appl. Phys. **28**, L2112 (1989)
6. S. Nakamura, T. Mukai, M. Senoh, Appl. Phys. Lett. **64**, 1687 (1994)
7. C. Wetzel, D. Volm, B.K. Meyer, K. Pressel, S. Nilsson, E.N. Mokhov, P.G. Baranov, Appl. Phys. Lett. **65**, 1033 (1994)
8. P. Waltereit, O. Brandt, A. Trampert, H.T. Grahn, J. Menniger, M. Ramsteiner, M. Reiche, K.H. Ploog, Nature **406**, 865 (2000)
9. A. Kobayashi, S. Kawano, Y. Kawaguchi, J. Ohta, H. Fujioka, Appl. Phys. Lett. **90**, 041908 (2007)
10. Y. Daigo, N. Mutsukura, Thin Solid Films **483**, 38 (2005)
11. S.S. Lee, S.M. Park, P.J. Chong, J. Mater. Chem. **3**, 347 (1993)
12. B. Zhou, X. Li, T.L. Tansley, K.S.A. Butcher, J. Cryst. Growth **160**, 201 (1996)
13. S. Yamazaki, T. Yatsui, M. Ohtsu, T.-W. Kim, H. Fujioka, Appl. Phys. Lett. **85**, 3059 (2004)
14. S. Yamazaki, T. Yatsui, M. Ohtsu, Appl. Phys. Exp. **1**, 061102 (2008)
15. S. Yamazaki, T. Yatsui, M. Ohtsu, Appl. Phys. Exp. **2**, 031004 (2009)
16. H. Okabe, M.K. Emadi-Babaki, V.R. McCrary, J. Appl. Phys. **69**, 1730 (1991)
17. K. Watanabe, J. Chem. Phys. **22**, 19 (1954)
18. Y. Kawakami, Z. Gang Peng, Y. Narukawa, S. Fujita, S. Fujita, S. Nakamura, Appl. Phys. Lett. **69**, 1414 (1996)
19. S. Chichibu, T. Azuhata, T. Sota, S. Nakamura, J. Appl. Phys. **79**, 2784 (1996)
20. W. Shan, T.J. Schmidt, X.H. Yang, S.J. Hwang, J.J. Song, Appl. Phys. Lett. **66**, 985 (1995)
21. Y.P. Varshni, Physica **34**, 149 (1967)
22. B. Monemar, J.P. Bergman, T. Lumdstrom, C.I. Harris, H. Amano, I. Akasaki, T. Detchprohm, K. Hiramatsu, N. Sawaki, Solid-State Electron. **34**, 41 (1997)
23. D.G. Zhao, S.J. Xu, M.H. Xie, S.Y. Tong, H. Yang, Appl. Phys. Lett. **83**, 677 (2003)
24. T. Kobayashi, A. Hashimoto, A. Yamamoto, Tech. Rep. IEICE **104** (2004) [in Japanese]
25. B.K. Ghosh, T. Tanikawa, A. Hashimoto, A. Yamamoto, Y. Ito, J. Cryst. Growth **49**, 422 (2003)
26. M. Leroux, N. Grandjean, M. Laugt, J. Massies, Phys. Rev. B **58**, R13371 (1998)
27. N. Grandjean, J. Massies, I. Grzegory, S. Porowski, Semicond. Sci. Technol. **58**, 16 (2001)
28. T.A. Witten, L.M. Sander, Phys. Rev. Lett. **47**, 1400 (1981)
29. T.A. Witten, L.M. Sander, Phys. Rev. B **27**, 5686 (1983)
30. Gy. Radnoczi, Phys. Rev. A **35**, 4012 (1987)
31. B. Mandelbrot, *The Fractal Geometry of Nature* (Freeman, San Francisco, 1982)
32. J. Nittmann, G. Daccord, H.E. Stanley, Nature **314**, 141 (1985)
33. E. Ben-Jacob, R. Godbey, N. Goldenfeld, J. Koplik, H. Levine, L. Sander, Phys. Rev. Lett. **55**, 1315 (1985)
34. J.D. Chen, D. Wilkinson, Phys. Rev. Lett. **55**, 1892 (1985)
35. K.J. Maloy, J. Feder, T. Jossang, Phys. Rev. Lett. **55**, 2688 (1985)
36. M. Matushita, M. Sano, Y. Hayakawa, H. Honjo, Y. Sawada, Phys. Rev. Lett. **55**, 53 (1984)
37. D. Grier, L.M. Sander, R. Clarke, E. Ben-Jacob, Phys. Rev. Lett. **56**, 1264 (1986)
38. Y. Sawada, A. Dougherty, J.P. Gollub, Phys. Rev. Lett. **56**, 1260 (1986)
39. R. Brady, R.C. Ball, Nature **309**, 225 (1984)
40. W.T. Elam, S.A. Wolf, J. Sprague, D.V. Gubser, D. Van Vechten, G.L. Barz, P. Meakin, Phys. Rev. Lett. **54**, 701 (1985)
41. R. He, X. Qian, J. Yin, Z. Zhu, Chem. Phys. Lett. **369**, 454 (2003)
42. R. Kurimoto, A. Kawaguchi, J. Macromol. Sci. Phys. **42**, 441 (2003)
43. T. Ishikawa, S. Kohmoto, K. Asakawa, Appl. Phys. Lett. **73**, 1712 (1998)
44. S. Kohmoto, H. Nakamura, T. Ishikawa, K. Asakawa, Appl. Phys. Lett. **75**, 3488 (1999)
45. M. Ara, H. Graaf, H. Tada, Appl. Phys. Lett. **80**, 2565 (2002)
46. V.V. Polonski, Y. Yamamoto, M. Kourogi, H. Fukuda, M. Ohtsu, J. Microscopy **194**, 545 (1999)
47. Y. Yamamoto, M. Kourogi, M. Ohtsu, V. Polonski, G.H. Lee, Appl. Phys. Lett. **76**, 2173 (2000)

48. T. Yatsui, T. Kawazoe, M. Ueda, Y. Yamamoto, M. Kourogi, M. Ohtsu, Appl. Phys. Lett. **81** 3651 (2002)

49. T. Yatsui, S. Takubo, J. Lim, W. Nomura, M. Kourogi, M. Ohtsu, Appl. Phys. Lett. **83**, 1716 (2003)

50. J. Lim, T. Yatsui, M. Ohtsu, IEICE Trans. Electron. **E-88C**, 1832 (2005)

51. Y. Yamamoto, M. Kourogi, M. Ohtsu, G.H. Lee, T. Kawazoe, IEICE Trans. Electron. **E85-C**, 2081 (2002)

52. M. Ohtsu (ed.), *Near-field Nano/Atom Optics and Technology* (Springer, Tokyo, 1999)

53. K. Karrai, R.D. Grober, Appl. Phys. Lett. **66**, 1842 (1995)

54. A. Dräbenstedt, J. Wrachtrup, C. von Borczyskowski, Appl. Phys. Lett. **68**, 3497 (1996)

55. F.J. Giessibl, Appl. Phys. Lett. **76**, 1470 (2000)

56. S. Cho, J. Ma, Y. Kim, Y. Sun, G. Wong, J.B. Ketterson, Appl. Phys. Lett. **75**, 2761 (1999)

57. T. Aoki, Y. Hatanaka, D.C. Look, Appl. Phys. Lett. **76**, 3275 (2000)

58. R.R. Krchnavek, H.H. Gilgen, J.C. Chen, P.S. Shaw, T.J. Licata, R.M. Osgood, Jr., J. Vac. Sci. Technol. B **5**, 20 (1987)

59. R.L. Jackson, J. Chem. Phys. **96**, 5938 (1992)

60. C.J. Chen, R.M. Osgood, Jr., Chem. Phys. Lett. **98**, 363 (1983)

61. G.T. Boyd, Th. Rasing, J.R.R. Leite, Y.R. Shen, Phys. Rev. B **30**, 519 (1984)

62. A. Wokaum, J.P. Gordon, P.F. Liao, Phys. Rev. Lett. **48**, 957 (1982)

63. K.F. MacDonald, V.A. Fedotov, S. Pochon, K.J. Ross, G.C. Stevens, N.I. Zheludev, W.S. Brocklesby, V.I. Emel'yanov, Appl. Phys. Lett. **80**, 1643 (2002)

64. J. Bosbach, D. Martin, F. Stietz, T. Wenzel, F. Trager, Appl. Phys. Lett. **74**, 2605 (1999)

65. H. Kuwata, H. Tamaru, K. Miyano, Appl. Phys. Lett. **83**, 4625 (2003)

66. R.G. Yarovaya, I.N. Shklyarevsklii, A.F.A. El-Shazly, Sov. Phys.-JETP **38**, 331 (1974)

67. R.L. Jackson, Chem. Phys. Lett. **163**, 315 (1989)

68. A. Sato, Y. Tanaka, M. Tsunekawa. M. Kobayashi, H. Sato, J. Phys. Chem. **97**, 8458 (1993)

69. P.J. Young, R.K. Gosavi, J. Connor, O.P. Strausz, H.E. Gunning, J. Chem. Phys. **58**, 5280 (1973)

70. M. Ohtsu, K. Kobayashi (eds.) *Optical Near Fields* (Springer, Berlin, 2003)

71. R.G. Yarovaya, I.N. Shklyarevsklii, A.F.A. El-Shazly: Sov. Phys. JETP **38** 331 (1974)

Chapter 3
Self-assembled Size Regulation and Its Alignment

3.1 Introduction

Optical near-field energy transfer between quantum dots (QDs) has been applied to the production of nanophotonic devices [1–4] because it decreases the device size beyond the diffraction limit of light and achieves novel functions unattainable using conventional propagating light. Logic devices, including AND gates [5] and NOT gates [6], light-harvesting nanofountains [7], and nanophotonic couplers [8] have been demonstrated as nanophotonic devices. Several semiconductor materials, including [9], InAs [10], InGaAs [11], and CdCe [8], have been used for QDs. Additionally, ZnO is a promising candidate material because of its large exciton binding energy [12]. Moreover, AND-gate operation has been demonstrated using ZnO nanorod multi-quantum wells [12].

For more advanced nanophotonic devices using ZnO, it is advantageous to replace the nanorod quantum wells with QDs for larger confinement energies. For this replacement, QD size should be controlled to ensure that the quantized energy levels are resonant, facilitating efficient optical near-field interaction. For ZnO QDs that are 5 nm, the size-mismatching between the QDs must be within 10% (i.e., 0.5 nm) to maintain the resonance condition, as estimated based on the broadening (30 meV) of the discrete energy levels of exciton, which is determined by the magnitude of the thermal energy at room temperature [3]. Among methods used to grow ZnO QDs, such as laser ablation [13] and molecular beam epitaxy [14], synthetic methods using liquid solutions are advantageous because they result in high productivity and size-controlled ZnO QDs [15]. However, size distributions of the QDs fluctuate depending on the thermal equilibrium condition of the chemical reaction [16], and the fluctuation can be as large as 25% for the conventional sol–gel method [2, 15].

This study introduced photo-induced desorption to the sol–gel method to reduce size fluctuations. When synthesized ZnO QDs are illuminated by light with photon energies higher than their bandgap energy, electron–hole pairs trigger an oxidation–reduction reaction in the QDs; thus, the ZnO atoms depositing on the QD surface

T. Yatsui, *Nanophotonic Fabrication*, Nano-Optics and Nanophotonics,
DOI 10.1007/978-3-642-24172-7_3, © Springer-Verlag Berlin Heidelberg 2012

are desorbed. The growth rate is controlled by the absorbed light intensity and wavelength, which control the QD size. Related methods for size-controlled QDs using photo-induced chemical processes have been reported for CdSe [17] and Si [18]. However, they were limited to etching the materials after growth, and particle size distributions were not quantitatively evaluated. This chapter discusses the role of photo-induced desorption in the sol–gel synthesis of ZnO QDs and the decrease in luminescence intensity due to the defect levels in ZnO QDs.

3.2   Size-, Position-, and Separation-Controlled One-Dimensional Alignment of Nanoparticles Using an Optical Near Field

Promising components for integrating the nanometer-sized photonic devices include chemically synthesized nanocrystals, such as metallic nanocrystals [19], semiconductor quantum dots [20], and nanorods [21], because they have uniform size, controlled shape, defined chemical composition, and tunable surface chemical functionality. However, position- and size-controlled deposition methods have not yet been developed. Since several methods have been developed to prepare nanometer-sized templates reproducibly [22], it is expected that the self-assembly of colloidal nanostructures into a lithographically patterned substrate will enable precise control at all scales [23]. Capillary forces play an important role, because colloidal nanostructures are synthesized in solution. Recently, successful integration of polymer or silica spheres [24, 25] and complex nanostructures such as nanotetrapods [25] into templates by controlling the capillary force using appropriate template structures has been reported, although their size and separation are typically uniform.

To fabricate nanophotonic devices, we propose a novel method of assembling nanoparticles by controlling the capillary force interaction and suspension flow. Further control of the positioning and separation of the nanoparticles is realized by controlling the particle–particle and particle–substrate interactions using an optical near field.

To control position and separation very accurately, preliminary experiment was performed on a patterned Si substrate, where an array of $10\,\mu$m holes in 100-nm-thick SiO_2 was fabricated using photolithography (Fig. 3.1a). Subsequently, a suspension containing latex beads with a mean diameter of 40 nm was dispersed on the substrate and the latex beads were aligned after solvent evaporation. The deposited latex beads were not subjected to any surface treatment and were dispersed in pure water at 0.001 wt%. Although the 10-μm-sized template resulted in low selectivity in the position of the latex beads (Fig. 3.1b, c), the beads were deposited only on the SiO_2 surface owing to its higher capillarity.

For higher positional selectivity, the suspension containing latex beads was dropped onto a lithographically patterned Si substrate that was spinning at 3,000 revolutions per minute (rpm). As shown in Fig. 3.2a, the suspension flow split into

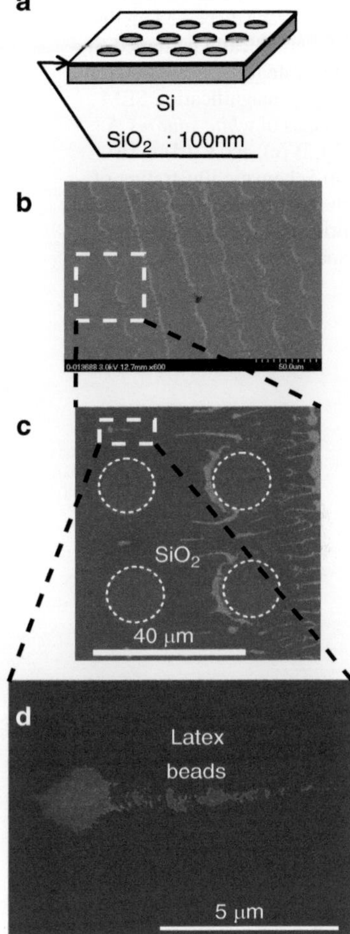

Fig. 3.1 (**a**) Schematic of lithographically patterned Si substrate. (**b–d**) SEM images of latex beads dispersed on the lithographically patterned Si substrate

two branches at the SiO$_2$ hole. SEM images (Fig. 3.2b–d) show that the chain of colloidal beads was aligned at the Si/SiO$_2$ interface. Note that the number of rows of latex beads decreased (Fig. 3.2b, c) and only the smallest beads, which were 20 nm in diameter, reached the end of the suspension flow (Fig. 3.2d). Assuming the same particle-suspension contact angle (denoted ψ in Fig. 3.2e) for various particles diameters, the flow speed of the larger latex beads had greater deceleration since the magnitude of the force pushing the particles on the SiO$_2$ (denoted F in Fig. 3.2e) owing to evaporation of the solvent is proportional to the particle diameter [25]. In other ward, the size selection was realized.

Based on the results of preliminary deposition, we tried assembling metallic nanoparticles because they are the material used to construct nano-dot couplers [26]. In this trial, we investigated the assembly of colloidal gold nanoparticles with a mean diameter of 20 nm dispersed in citrate solution at 0.001%. The nanoparticles

Fig. 3.2 (a) SEM image of latex beads dispersed on the lithographically patterned Si substrate rotated at 3,000 rpm. Higher magnification SEM images of *white squares* A (b), B (c), and C (d) in (a). (e) Schematic illustrating of the particle-assembly process driven by the capillary force and suspension flow

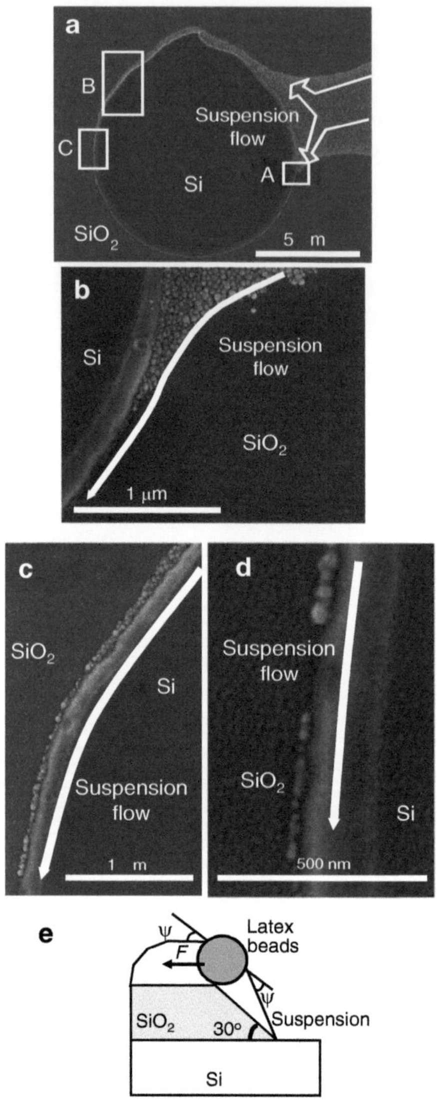

were prepared by the citric acid reduction of gold ions and terminated by a carboxyl group (approximate length is 0.2 nm) with a negative charge [26]). However, they could not be aggregated using the same deposition process as for the latex beads (Fig. 3.3a). To aggregate these particles, we fabricated a SiO$_2$ line structure with a plateau width of 50 nm on the Si substrate using photolithography (Fig. 3.3b). The solvent containing the colloidal gold nanoparticles was dropped onto this substrate at 3,000 rpm. Then, the colloidal gold nanoparticles aggregated along the plateau of the SiO$_2$ line (Fig. 3.3b, c). This indicates that the capillary force induced by

Fig. 3.3 (**a**) SEM image of colloidal gold nanoparticles dispersed on the lithographically patterned Si substrate rotated at 3,000 rpm. (**b**) Schematic of the SiO$_2$ line structure fabricated on the Si substrate. (**c**) SEM images of colloidal gold nanoparticles dispersed on the SiO$_2$ line rotated at 3,000 rpm. Inset; cross-section of the substrate along the *white line* (*dashes and dots*)

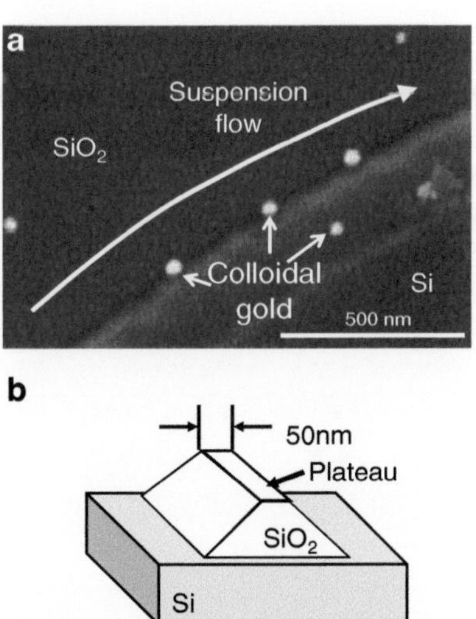

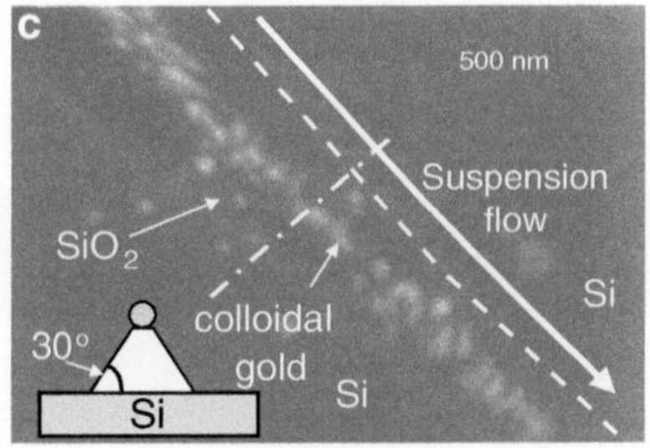

the lithographically patterned substrate, which is caused by the higher wettability of SiO$_2$ than that of the Si, was larger than the repulsive force owing to the negative charge of the carboxyl group on the colloidal gold nanoparticles, and this resulted in the aggregation and alignment of the colloidal gold nanoparticles at high density. This indicates that the capillary force induced by the lithographically patterned substrate, which is caused by the higher wettability of SiO$_2$ than that of the Si, was larger than the repulsive force owing to the negative charge of the carboxyl group on the colloidal gold nanoparticles, and this resulted in the aggregation and alignment of the colloidal gold nanoparticles at high density (Fig. 3.4).

To further control size, separation, and positioning, we examined the aggregation of colloidal gold nanoparticles under illumination, because the colloidal gold nanoparticles have strong optical absorption. Strong absorption should desorb the carboxyl group from the colloidal gold nanoparticles and result in their aggregation. Such an aggregation of colloidal gold nanoparticles were confirmed by the illumination of light. Figure 3.5a shows the aggregated gold nanoparticles over the pyramidal Si substrate under the 690-nm-light illumination for 60 s. However, since the light was illuminated through the droplet of the colloidal gold nanoparticles, aggregated colloidal gold nanoparticles were spread outside the beam spot.

Fig. 3.4 (**a**) Frontal
illumination. (**b**) Aggregated
colloidal gold nanoparticles
with frontal illumination
under 690-nm light
(25 mW/mm^2) for 60 s

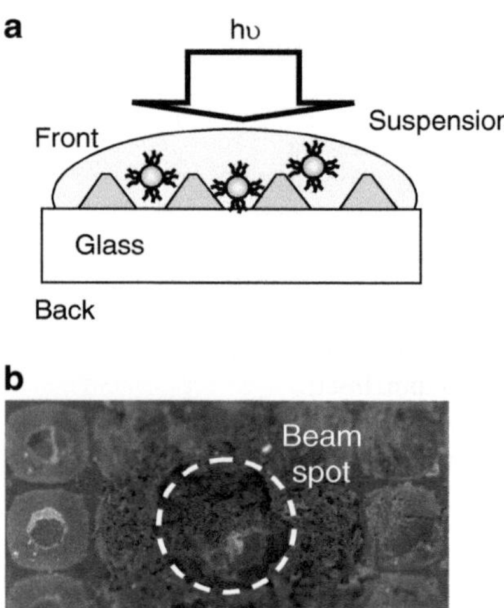

Fig. 3.5 (**a**) Schematic of the
experimental setup. (**b**) SEM
image of the fabricated Si
wedge structure

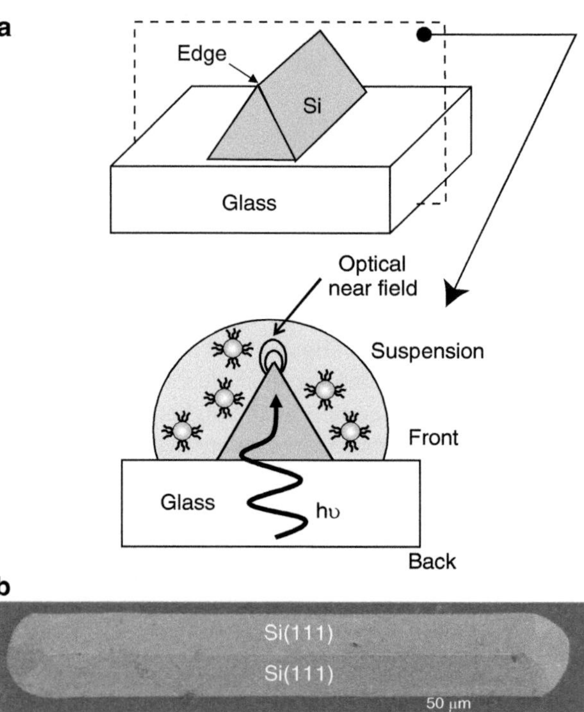

In order to realize selective aggregation of the gold nanoparticles at the desired position, the suspension was illuminated from behind (Fig. 3.5b). Furthermore, we used a Si wedge, because this is a suitable structure for a far-/near-field conversion device [27]. The Si wedge structure (Fig. 3.5c) was fabricated by the photolithography and anisotropical etching of Si. Detailed fabrication process is described in Fig. 3.6.

For this structure, colloidal gold nanoparticles were deposited around the edge after evaporating the suspension without illumination (Fig. 3.7a, b). Such aggregation is owing to its wedge structure. This is because the suspension at the edge is thinner than that on the Si(111) plane owing to its low capillarity, and this causes the convective transport of particles toward the edge [28]. Further selective alignment along the edge of the Si wedge was realized using rear illumination. Figure 3.7c, d shows the deposited colloidal gold nanoparticles with illumination under 690-nm light (25 mW/mm^2) for 60 s. Since the optical near-field energy is enhanced at the edge owing to the high refractive index of Si (see Fig. 3.5b) [16], selective aggregation along the edge with higher density is seen in these figures. This is due to the desorption of the carboxyl group by the absorption of light by the colloidal gold nanoparticles.

Note that the colloidal gold nanoparticles were closely aggregated and aligned linearly to form a wire shape when the polarization was perpendicular to the edge axis (Fig. 3.7c), while they were aligned with separation of several tens of nanometers in the parallel polarization (Fig. 3.7d]). As the optical near-field energy for parallel polarization is higher than that for perpendicular polarization [27], greater aggregation is expected for parallel polarization. Nevertheless, the parallel polarization resulted in less aggregation. The low resolution of SEM images does

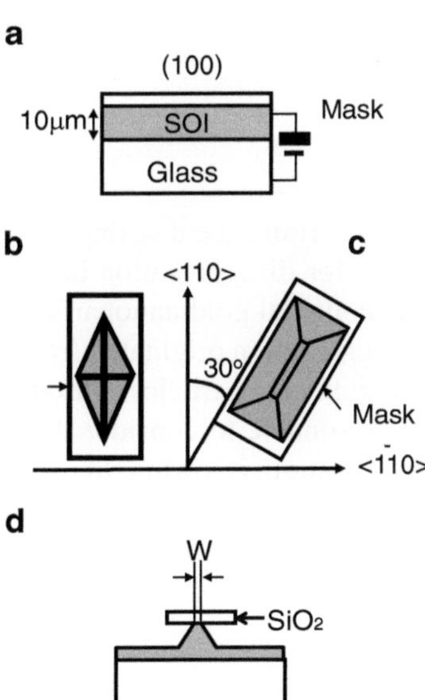

Fig. 3.6 Fabrication steps for a pyramidal Si probe: (**a**) anodic bonding; (**b**), (**c**), (**d**) anisotropic etching for fabrication of the Si wedge

Fig. 3.7 (**a**) Overview of the Si wedge structure. (**b**) SEM image of colloidal gold nanoparticles deposited on the edge of the Si wedge structure without illumination. SEM images of colloidal gold nanoparticles on the Si wedge structure under illumination with polarization perpendicular (**c**) and parallel (**d**) to the edge. Schematic diagrams of the aggregation of colloidal gold nanoparticles along the edge of the Si wedge with polarization perpendicular (**e**) and parallel (**f**) to the edge

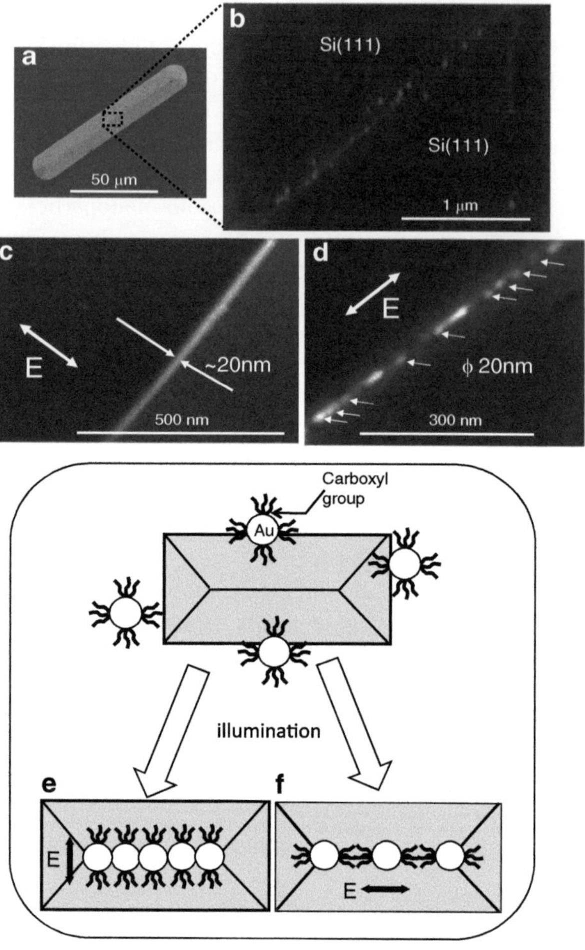

not determine the distribution of the carboxyl molecules. However, such a repulsive force for disaggregation is caused by the carboxyl molecules which remained on the colloidal gold nanoparticles. Thus, we believe that the difference in the degree of aggregation originated from differences in the charge distribution induced inside the gold nanoparticles. Based on the polarization dependence of the aggregation, it is reasonable to consider that the aggregation along the edge with perpendicular polarization is owing to partially adsorbed carboxyl groups (Fig. 3.7e), while the disaggregation with the parallel polarization resulted from the repulsive force induced by the partially attached carboxyl group on the colloidal gold nanoparticles (Fig. 3.7f).

3.3 Self-assembly of Size- and Position-Controlled Ultra-long Nanodot Chains Using Near-Field Optical Desorption

Based on the technique described in the Sect. 2.2.3, we propose a self-assembling method that builds nanodot chains by controlling the desorption with an optical near-field. Our approach is illustrated schematically in Fig. 3.8a. A nanodot chain of metallic nanoparticles was fabricated using radio frequency (RF) sputtering under illumination on glass substrate. In order to realize self-assembly, a simple groove 100-nm wide and 30-nm deep was fabricated on the glass substrate. During deposition of the metal, linear polarized light illuminating the groove directly above (E_{90}) was used to excite a strong optical near-field at the edge of the groove (see Fig. 3.8b), which induced the desorption of the deposited metallic nanoparticles [29]. A metallic dot has a strong optical absorption due to plasmon resonance [30, 31], which strongly depends on particle size. This can induce desorption of the deposited metallic nanodot when it reaches the resonant diameter [32, 33]. As the deposition of metallic dots proceeds, the growth is governed by a trade-off between deposition and desorption, which determines dot size, depending on the photon energy of the incident light. Consequently, the metallic nanoparticles should align along the groove (Fig. 3.8b, c).

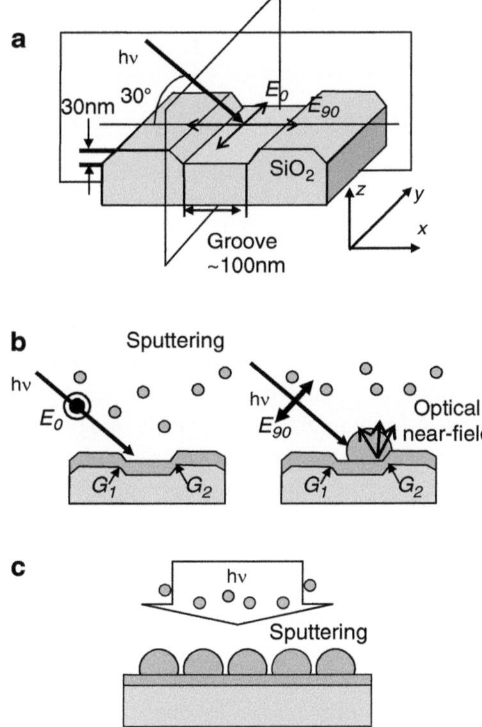

Fig. 3.8 (**a**) Size- and position-controlled ultra-long nanodot chain formation. The slanted light had a spot diameter of 1 mm. The groove parallels axis y. E_{90} and E_0 are perpendicular and parallel to axis y, respectively. (**b**), (**c**) Cross-sections in planes XZ and YZ, respectively

Fig. 3.9 (**a**), (**b**) SEM images of deposited aluminum with perpendicular polarization E_{90} ($h\nu = 2.33$ eV). (**c**), (**d**) Respective AFM images of the surface of the glass substrate before and after aluminum deposition, at the same position. (**e**) Curves A and B show the respective cross-sectional profiles of AFM images along *dashed white lines* A in (**c**) and B in (**d**). Scale bars: (**a**) 500 nm, (**b**) 10 μm

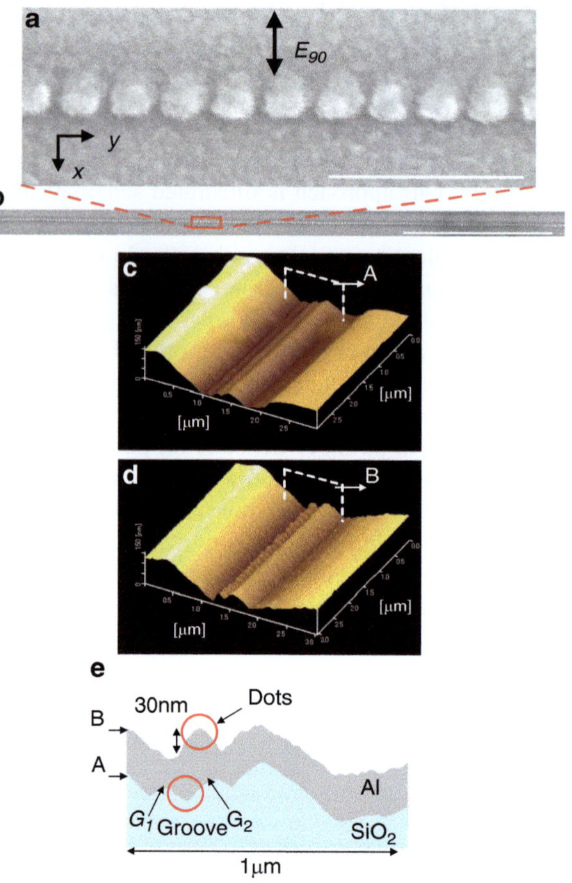

Illumination with 2.33-eV light (50 mW) during the deposition of aluminum resulted in the formation of 99.6-nm-diameter aluminum nanodot chains with 27.9-nm separation that were as long as 100 μm in a highly size- and position-controlled manner (see Figs. 3.8b and 3.9a). The deviations of both nanodot size and the separation, determined from scanning electron microscopy (SEM) images, were as small as 5 nm. To identify the position of the chain, we compared topographic atomic force microscopy (AFM) images of the surface of the glass substrate before (Fig. 3.9c) and after (Fig. 3.9d) aluminum deposition, at the same position. Curves α and β in Fig. 3.9e show the respective cross-sectional profiles along dashed white lines a–a' in Fig. 3.9c and b–b' in Fig. 3.9d. Comparison of these profiles showed that the nanodot chain formed around edge G_2. Furthermore, illumination with parallel polarization along groove E_0 resulted in film growth along the groove structure and no dot structure was obtained. Since the near-field intensity with E_{90} was strongly enhanced at the metallic edge of the groove in comparison with E_0 owing to edge enhancement of the electrical field (see Fig. 3.8b), a strong near-field intensity results in nanodot chain formation.

Fig. 3.10 The field intensities ($|E|^2$) calculated using the FDTD method for the bare glass substrate (**a, c**) and glass coated with 20-nm-thick aluminum film (**b, d**). Scale bars: (**c**), (**d**) 100 nm

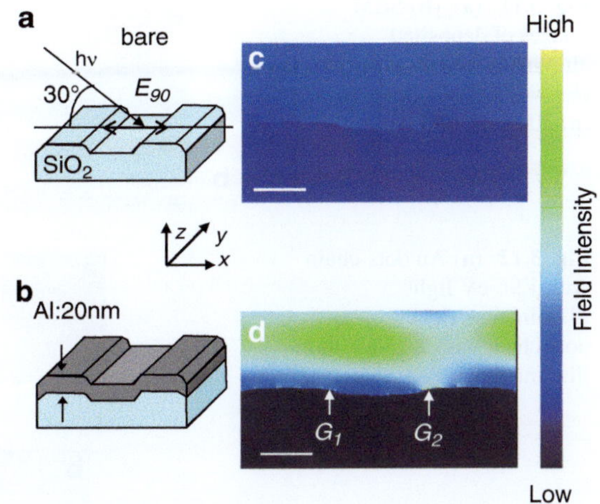

Dot formation at the one-sided edge originates from the asymmetric electric-field intensity distribution, owing to the slanted illumination. To confirm this prediction, we calculated the electric field intensity using finite-difference time-domain (FDTD) method (Fig. 3.10a, b). We compared the electric field intensity distribution between a bare glass groove (100-nm wide and 30-nm deep) and one coated with 20-nm-thick aluminum ($n = 0.867 + i6.42$) with slanted (30°) illumination at a photon energy of 2.33 eV. Although the bare groove in the glass does not induce localization of the electric field intensity (Fig. 3.10c), we found that the slanted illumination strongly enhanced the electric field intensity at the back side of metallized groove G_2 (Fig. 3.10d) only, indicating desorption of the metallic nanoparticles. It is reasonable to assume that the desorption was excited after the deposition of the aluminum film. The oscillations in the electric field perpendicular to the edge induce charges on the metallized edge of the groove, and these induced charges generate a strong electric field and the associated scattering [34].

Aluminum-dot chain formation was also observed with RF sputtering of aluminum under illumination with 2.62-eV light (100 mW) with E_{90} using the same grooved (100-nm wide and 30-nm deep) glass substrate, which resulted in the formation of 84.2-nm nanodots with 48.6-nm separation (see Fig. 3.11a, b). Although the deviations of both nanodot size and the separation were as large as 10 nm, the dot size was reduced in proportion to the increase in the photon energy (99.6 nm × (2.33/2.62) = 88.5 nm ∼ 84.2 nm). This result indicates that the obtained size is determined by the photon energy and that the size-controlled dot-chain formation originates from photo-desorption of the deposited metallic nanoparticles [29]. The period under 2.62-eV light illumination (132.8 nm) was longer than that (127.5 nm) for fabrication using the 2.33-eV light. However, the ratios of the center–center distance (d) and radius (a) of the nanodots ($d/a = 2.56$ and 3.15 obtained under 2.33-eV and 2.62-eV light illumination, respectively) are similar to the optimum value, which is in the range of 2.4–3.0, for the efficient transmission of the optical

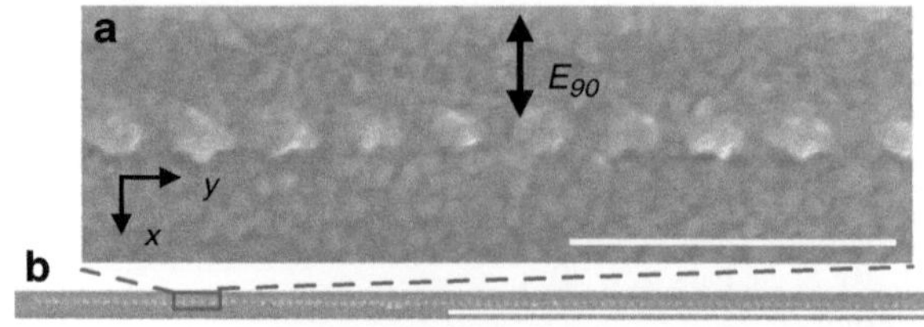

Fig. 3.11 (**a**), (**b**) SEM images of deposited aluminum with E_{90} ($h\nu = 2.62\,\text{eV}$). Scale bars: (**a**) 500 nm, (**b**) 5 μm

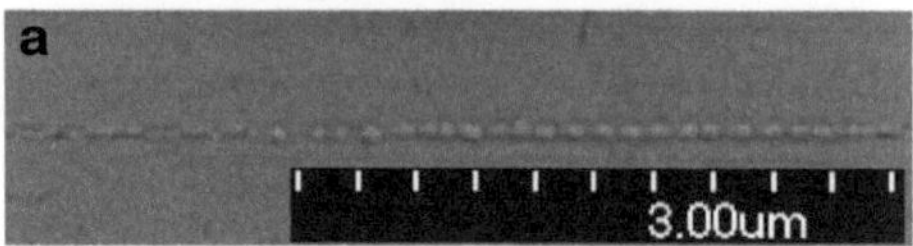

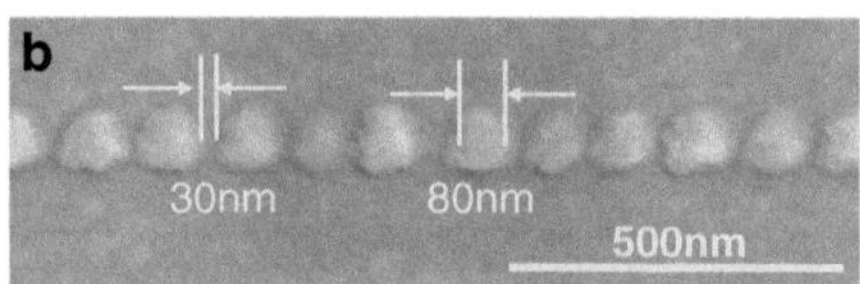

Fig. 3.12 (**a**) Au dots-chain with 1.96-eV light illumination. (**b**) Pt dots-chain with 2.33-eV light illumination

energy along the chain of spherical metal dots calculated using Mie's theory [35], which is determined by the trade-off between the increase in the transmission loss in the metal and the reduction in the coupling loss between adjacent metallic nanoparticles as the separation increases. Future theoretical analysis, which will include the effect of a continuous metallic film below the nanodot chain, is required to explain the optimum separation of the nanoparticles depending on the photon energy. However, our preliminary results imply that the center–center distance is set at the optimum distance for efficient energy transfer of the optical near-field, given that such a strong optical near-field can induce desorption of the deposited metallic nanoparticles and result in position-controlled dot-chain formation.

This technique was also applicable to other metal. By illuminating 1.96 eV (2.33 eV) light during the deposition of Au (Pt) film, we successfully fabricated Au (Pt) dots chain as long as 20 μm (see Fig. 3.12a, b). In addition, this technique was applied to the semiconductor. Figure 3.13 shows the results of GaN. By illuminating 3.81 eV light during the deposition of GaN, the drastic reduction of size deviation was confirmed (see Fig. 3.13d).

We believe that the resulting structure should have high transmission efficiency for optical near-field energy, which is suitable for a nanodot coupler, given that the nanodot chain forms at a size based on plasmon resonance, and we have already reported the efficient energy transfer along a nanodot chain involving a continuous metallic film below the nanodot chain [36, 37]. Furthermore, since our deposition method is based on a photo-desorption reaction, illumination at multiple photon energies using the simple lithographically patterned substrate could realize the simultaneous fabrication of size- and position-controlled nanoscale structures with different particle sizes, and with other metals or semiconductors. The use

Fig. 3.13 (**a**) Sputtered GaN nanoparticles (**a**) without illumination and (**b** and **c**) with illumination. (**d**) Histogram of the size of GaN nanoparticles. Curve A and B represent of (**a**) and (**b**). Curves A and B were fitted by 16 ± 12 nm (A) and 21 ± 8 nm

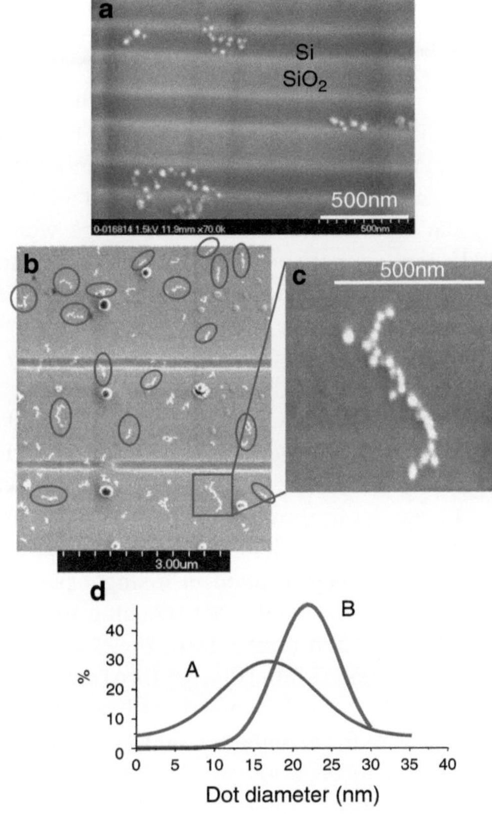

of the self-assembling method with a simple lithographically patterned substrate will dramatically increase the throughput of the production of nanoscale structures, required by future systems.

3.4   High-Resolution Capability of Optical Near-Field Imprint Lithography

As next-generation lithography (NGL) for the 32-nm node and below, ArF immersion lithography, extreme ultraviolet lithography (EUVL), and electron projection lithography (EPL) have been studied. However, a practical problem is the increasing cost and size of NGL tools. To solve these problems, optical near-field lithography has been developed by introducing tri-layer resist. The fabrication of sub-50 nm features has been realized using the i-line ($\lambda = 365$ nm) [38], with conventional photolithography facilities. A further decrease in feature size has been reported with the introduction of imprint lithography [39], which resulted in the fabrication of 14-nm pitch features [40]. Although conventional imprint lithography results in the same mold structure, the use of the optical near-field intensity distribution should realize features smaller than the mold structure, since a nanoscale mold has a nanoscale optical field distribution at its edge. In this study, we propose and demonstrate optical near-field imprint lithography to introduce its ability to obtain a higher resolution than the size of the mold features.

To realize the efficient excitation of an optical near-field on a mold, the thickness of the metallic film and the coverage were optimized using the FDTD method [41]. For comparison with a conventional photolithographic mask, the optical field distribution for 80-nm half-pitch, 200-nm-deep Cr line-and-space (LS) on the SiO_2 ($n = 1.5$) substrate was calculated at a wavelength of 436 nm (g-line). Here, the refractive index of Cr was assumed to be $n = 1.78 + i2.695$, and the line was parallel to the y-axis (Fig. 3.14a). Since the imprint mold used in this study was fabricated using SiO_2, we also calculated the optical field distribution of 80-nm half-pitch, 200-nm-deep SiO_2 LS, coated with aluminum film ($n = 0.56 + i5.2$) [42] (Fig. 3.14b–d). The minimum cell size was $5 \times 12.5 \times 5$ nm.

Figure 3.14e shows the cross-sectional profile along the x-axis 10 nm from the mold. The optical field intensity distribution of the Cr LS used for conventional photolithography resulted in a single peak corresponding to the space of the Cr mask, which resulted from reducing the optical field intensity through the 200-nm-thick Cr film (curve 1a in Fig. 3.14e). By contrast, coating the SiO_2 LS with a 20-nm-thick Al film (Fig. 3.14b, c) enhanced the electric field intensity at the edge of the mold (curves 2a and 3a in Fig. 3.14e). Higher localization at the edge of the mold was realized without a sidewall coating (Fig. 3.14b and curve 2a in Fig. 3.14e). Since this localization was not observed for the thicker coating (a 50-nm-thick Al film on top of SiO_2 and a 20-nm-thick Al film on the sidewall) and was not observed in the y-polarization (curve y in Fig. 3.14f), this localized optical near-field originated from the edge effect. Since efficient excitation of the surface plasmon is obtained with a 15-nm-thick aluminum coating on glass in the Kretschmann configuration [43], this size-dependent feature is attributed to localized surface plasmon resonance on the Al film. Furthermore, the localization of the optical field intensity to an area as narrow as 25 nm in curve 2a in Fig. 3.14e infers the realization of a resolution higher than the mold size. This localization was also observed in curve 2b (Fig. 3.14e) obtained in the plane along the x-axis 10 nm from the bottom of the mold.

We performed imprint lithography to confirm the higher resolution capability using an optical near-field, as discussed above. Commercial photocurable acryl PAK01 resin (blended by Toyo Gosei) was used; it is composed of tri-propylene-glycol-diacrylate monomer with dimethoxy-phenyl-acetophenon as the photo-initiator and has good release properties [44]. Polycarbonate (PC) substrate was spin-coated with PAK01. We used 300-nm half-pitch, 200-nm-deep SiO_2 LS as the mold.

To obtain the optimum structure shown in Fig. 3.14b (20-nm-thick Al film with no sidewall coating), the mold was coated with Al using vacuum evaporation (Fig. 3.15a). The mold was pressed into the liquid polymer on PC substrate under

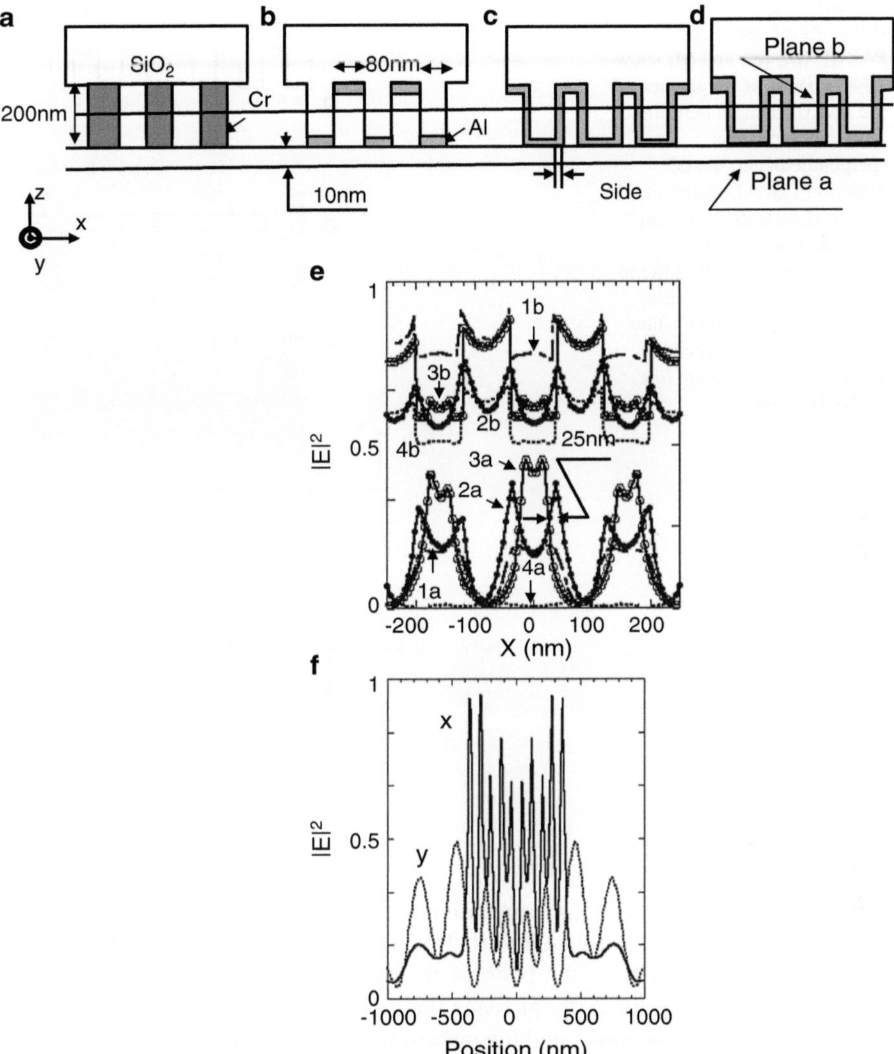

Fig. 3.14 Schematic of the calculation models. (a) Cr LS on SiO_2 substrate. (b–d) SiO_2 LS coated with (b) 20-nm Al without a sidewall coating, (c) 20-nm Al with a sidewall coating, and (d) 50-nm Al at the top and 20-nm Al on the sidewall. (e) Curves 1a–4a show the cross-sectional profiles along the x-axis, 10 nm from the mold (plane a) with five grooves in the mold. Curves 1b–4b show the cross-sectional profiles along the x-axis 10 nm from the bottom of the mold (plane b). The beam width at $1/e^2$ of incident light with a Gaussian shape was 1,000 nm. The x-coordinate is perpendicular to the grating in (a)–(d). (f) The polarization dependence of the cross-sectional profiles for x- (perpendicular to the LS direction) and y- (parallel to the LS direction) polarization for the mold with five grooves in the mold. The beam width at $1/e^2$ of incident light with a Gaussian shape was 2,000 nm ($1/e^2$)

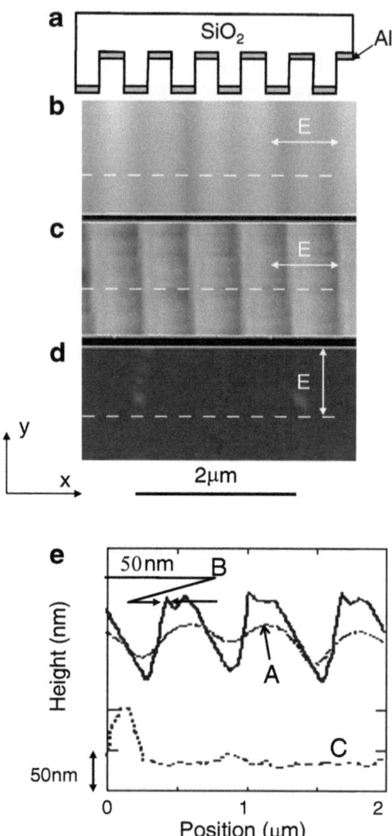

Fig. 3.15 (**a**) Mold position before its release from PAK01. (**b**), (**c**), and (**d**) show AFM images of the surface of the PAK01 using a bare mold with x-polarization (perpendicular to the LS direction), an Al-coated mold with x-polarization, and an Al-coated mold with y-polarization (parallel to the LS direction), respectively. (**e**) Curves A, B, and C show the cross-sectional profiles along the *dashed white lines* in (**b**), (**c**), and (**d**), respectively

a pressure of 70 kPa using a conventional contact mask aligner (MJB3, SUSS MicroTec KK). It was irradiated with UV light (g-line: 436 nm, power density: 30 mW/cm^2) for 30 s from the back of the mold, while maintaining the imprint pressure during exposure. After pressing the mold and UV curing, the PC substrate was separated from the mold, and the pattern was transferred.

First, we obtained topographic AFM images of the surface of PAK01 after release of the mold. Figure 3.15b–d shows the AFM images of a bare SiO$_2$ mold with x-polarization (perpendicular to the LS direction), an Al-coated SiO$_2$ mold with x-polarization, and an Al-coated SiO$_2$ mold with y-polarization (parallel to the LS direction), respectively. Although the bare mold resulted in a single pitch corresponding to the mold pitch (Fig. 3.15b, and curve A in Fig. 3.15e), we obtained sharp (50 nm) protruding structures at the edge of the Al-coated SiO$_2$ mold, when the mold was pressed under x-polarization (Fig. 3.15c and curve B in Fig. 3.15e). Although the pitch differs between the numerical and the experimental results, these profiles with protruding structures seen at the edge of the Al-coated SiO$_2$ mold are in good agreement with those calculated using FDTD (curve 2a of Fig. 3.14e). The calculated value at the point next to the interface is unstable in the FDTD calculation due to the drastic change in the refractive index. Furthermore, since the distribution of the optical near-field along the z-axis is as large as that along the x-axis, the optical near-field is believed to be localized to a region small as 20 nm along the z-axis. Therefore, we compared the profiles at the second point from the interface (10 nm apart from the interface). Although the calculated and obtained profiles differ, the protuberances were obtained only with x-polarization with an Al coating; therefore, we believe that the protuberances originated from the plasmon resonance, as predicted by the FDTD calculation. These results indicate that the resolution was higher than the pitch of the mold.

Next, we obtained scanning electron microscope (SEM) images of the transferred pattern (Fig. 3.16a–c). As shown in the AFM images, the SEM images confirmed that there were protuberances where the edge of the mold was pressed (inside the white solid ellipses in Fig. 3.16b.1, b.2) using the Al-coated mold with x-polarization.

The field distribution calculated using the FDTD method did not predict the resist structure after release of the mold. However, the correspondence between the AFM

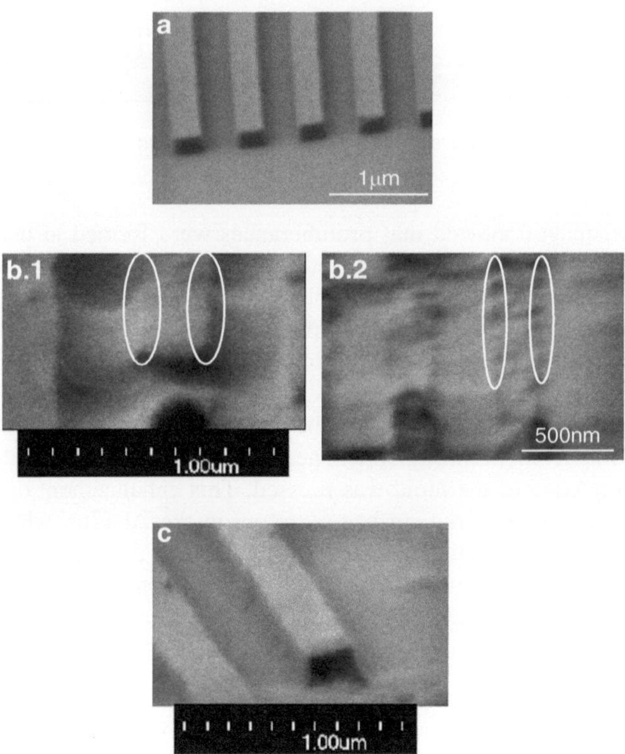

Fig. 3.16 Tilted (30°) SEM images using (**a**) a bare mold with x-polarization (perpendicular to the LS direction), (**b.1**) (**b.2**) an Al-coated mold with x-polarization, and (**c**) an Al-coated mold with y-polarization

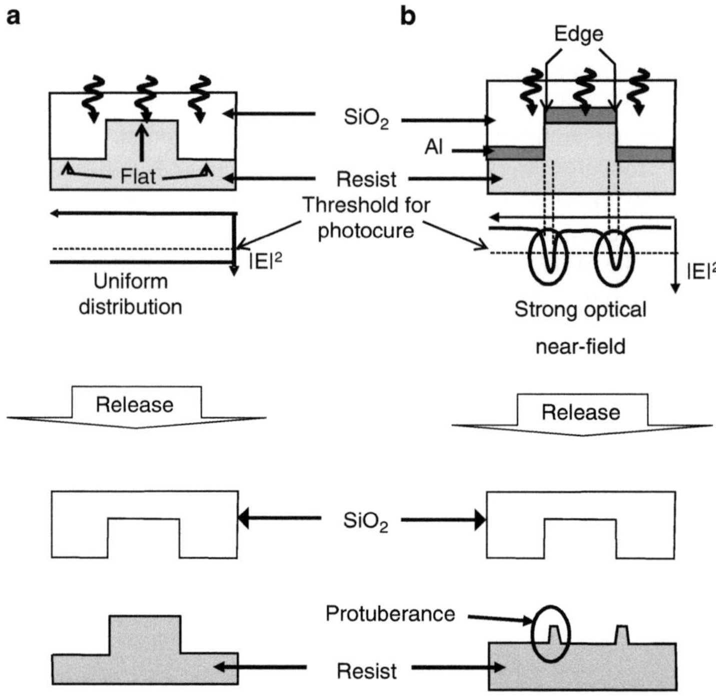

Fig. 3.17 Schematics of (**a**) conventional imprint lithography and (**b**) optical near-field imprint lithography

and the SEM images showed that protuberances were formed at positions where a strong optical near-field was localized. Based on these results, we concluded that using conventional imprint lithography, a strong uniform optical field intensity resulted in the formation of the same pitch as that of the mold (Fig. 3.17a). In contrast, with optical near-field imprint lithography, the resist was deformed in the underexposed condition where the flat surface of the mold was press, which was due to the Al coating of the mold, and the resist was remained in the overexposed condition arising from the strong optical near-field, which was due to the edge effect from where the edge of the mold was pressed. This enhancement originated from the resonant excitation of the surface plasmon on the Al film, which resulted in protuberances smaller than the pitch of the mold (see Fig. 3.17b). Future evaluations, which will include the effects of power and irradiation time, are required to explain the optimum dose for the higher contrast of the protuberances.

3.5 Self-assembly of ZnO QDs

For the sol–gel synthesis of ZnO QDs, 1.1 g of zinc acetate dihydrate and 0.29 g of lithium hydroxide monohydrate were dissolved in ethanol (50 mL) individually, and then the two solutions were mixed at 273 K [2, 16]. The mixed solution was kept

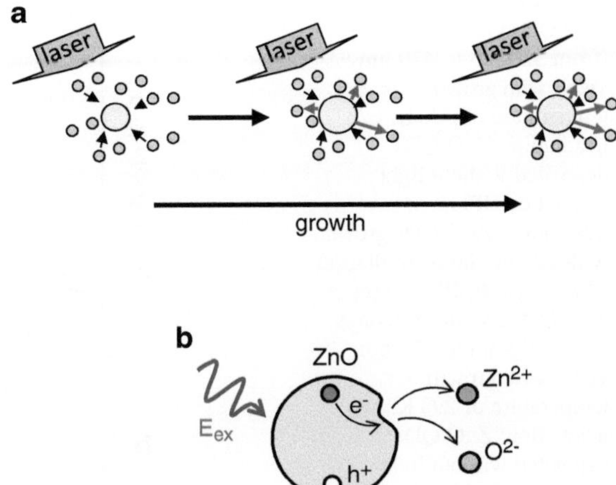

Fig. 3.18 (**a**) Schematic of laser-assisted sol–gel method. (**b**) Oxidation–reduction reaction triggered by the exciton excitation

at room temperature (300 K) to grow the ZnO QD particles. After the ZnO QDs grew to the desired size, the solution was mixed with 300 mL of hexane to remove the lithium, which was used to accelerate the growth. Then, the ZnO QD growth was halted; thus, ZnO QDs size was controlled. A continuum-wave (CW) He–Cd laser was used as a light source, and its wavelength ($\lambda = 325$ nm) was shorter than the absorption edge wavelength ($\lambda = 340$ nm). The irradiated laser (CW $\lambda = 325$ nm) having a uniform power density of 8 mW/cm³ over the mixed solution during whole growing process. The QD size was controlled by laser-assisted sol–gel method (see Fig. 3.18a). When synthesized ZnO QDs are illuminated by light with photon energies higher than their bandgap energy, electron–hole pairs trigger an oxidation–reduction reaction in the QDs; thus, the ZnO atoms depositing on the QD surface are desorbed (see Fig. 3.18b).

The synthesized ZnO QDs were dispersed uniformly over a sapphire substrate and were excited by the fourth harmonic ($\lambda = 266$ nm, power = 2 mW) of a YAG laser to measure their photoluminescence (PL) spectra at room temperature. The black and red curves in Fig. 3.19a show the measured profiles of QDs that were grown for 6 days either without light irradiation ($IPL_{w/o}$) or with irradiation from a 325-nm laser ($IPL_{w/325}$). The PL peak wavelength of $IPL_{w/325}$ was 6-nm blue-shifted compared with that of $IPL_{w/o}$, indicating that the growth rate decreased for the ZnO QDs under 325-nm laser irradiation.

To quantitatively analyze the mechanism of size control under light irradiation, we evaluated ZnO QD diameters. The shape of QDs synthesized by the sol–gel method is known to be spherical, although the crystalline structure of ZnO QDs is wurtzite (see Fig. 3.20) [45]. In the present study, the QD shape was assumed to be spherical. The QD diameter was calculated using the effective mass approximation, in which the emission energy E_{em} from the recombination of free exciton confined in QD can be describe by following equation [46]:

Fig. 3.19 (**a**) PL spectra of ZnO QDs grown for 6 days (from 320-nm to 420-nm view) with growth temperature of 300 K. w/o irradiation: ZnO QDs deposited without light irradiation ($IPL_{w/o}$); w/ irradiation: ZnO QDs grown with 325-nm laser irradiation ($PL_{w/325}$). (**b**) PL spectra of ZnO QDs grown for 6 days (from 300-nm to 600-nm view) with growth temperature of 293 K. w/o irradiation: ZnO QDs deposited without light irradiation; w/ irradiation: ZnO QDs grown with 325-nm laser irradiation

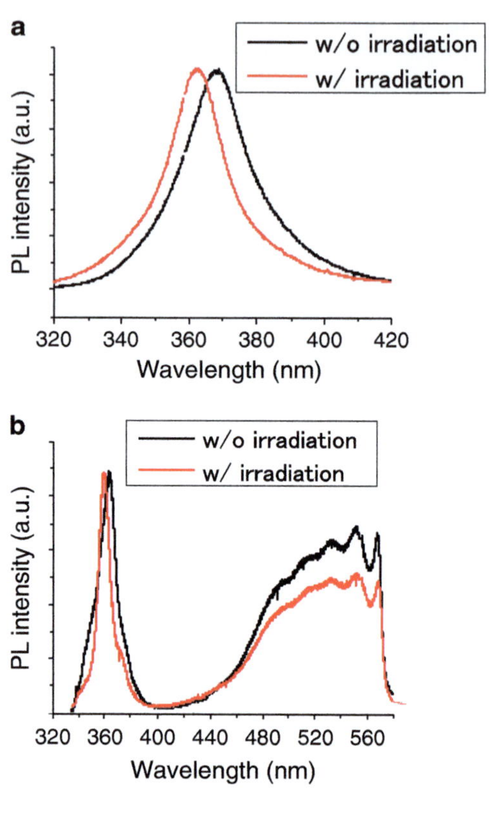

Fig. 3.20 Schematic of ZnO QD fabricated by sol–gel method

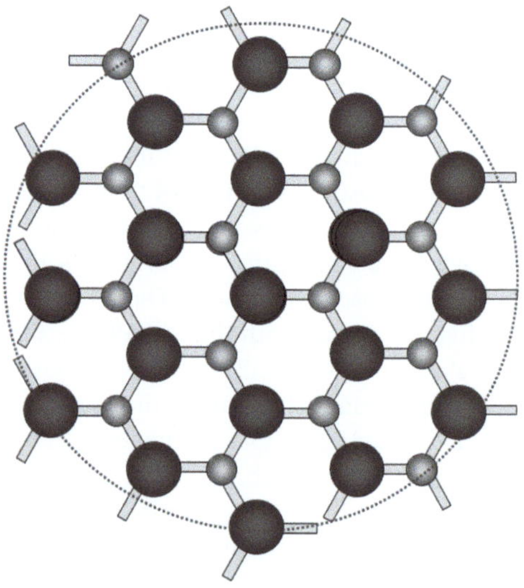

Fig. 3.21 (**a**) Relationship between QD diameter and PL peak wavelength. *Open circles* are plotted from average diameter determined using TEM images and the respective PL spectra. Using these obtained data, we determined m_e and m_h. The *black solid curve* are plotted using the values $m_e = 0.32\ m_0$, $m_h = 0.42\ m_0$ for (3.1). (**b**)–(**d**) TEM image of ZnO QD to determine diameter. (**e**) PL spectra of ZnO QDs. Curve B: PL spectrum of (**b**), curve C: PL spectrum of (**c**), and curve D: PL spectrum of (**d**)

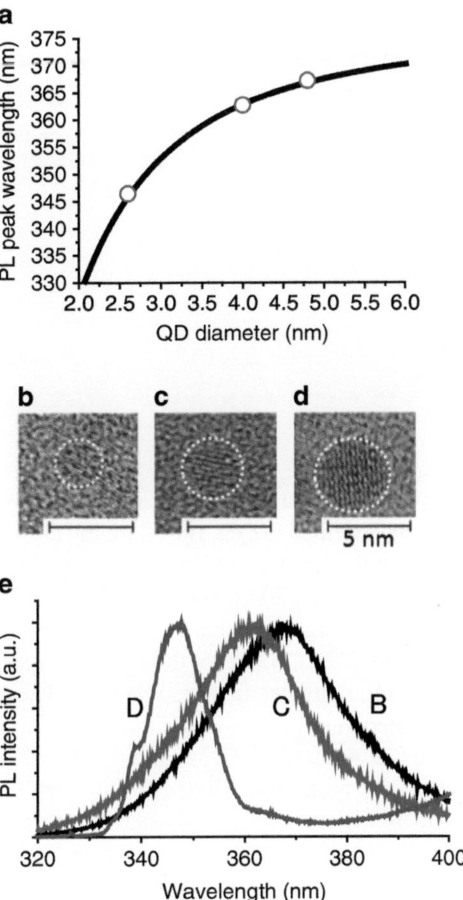

$$E_{em} = 3.37 + \frac{2\pi^2\hbar^2}{(m_e + m_h)\,ed^2} - \frac{13.6\mu}{m_0\varepsilon^2}. \tag{3.1}$$

Here d is the QD diameter, m_e and m_h the effective masses of electron and hole, m_0 the electron mass of 9.11×10^{-31} [kg], μ the translation mass described by $m_e m_h / (m_e + m_h)$, and e is the elementary electric charge. The third term in (3.1) is the correction term. The QD diameter d and the emission energy E_{em} were determined as average diameter determined using transmission electron microscopy (TEM) images which are shown in Fig. 3.21b–d, and the corresponding PL spectra which are shown in Fig. 3.21e, respectively, and they are plotted as open circles in Fig. 3.21a. Using these obtained data, we determined m_e and m_h. To obtain the black solid curve in Fig. 3.21 we use $m_e = 0.32\ m_0$, $m_h = 0.42\ m_0$. These values are in good agreement with the reported values in [47] ($m_e = 0.24\ m_0$ and $m_h = 0.45\ m_0$). Thus, we consider that the QD diameter can be determined using the effective

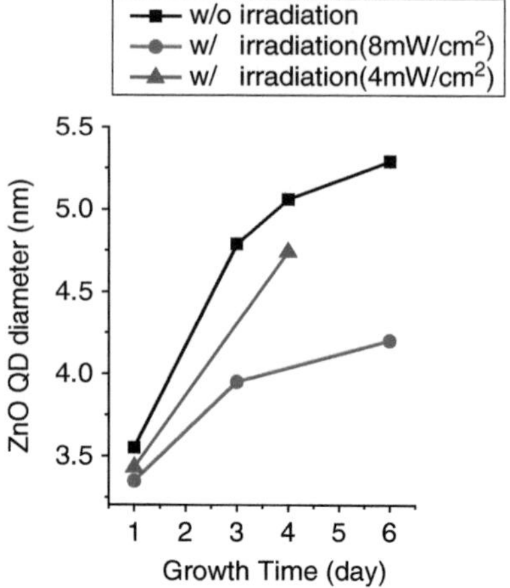

Fig. 3.22 Dependency of ZnO QD diameter on growth time

mass approximation and the emission is originated from the recombination of free exciton confined in QD. Based on the above estimation, we obtained the relationship between the growth time and the diameter (Fig. 3.22), showing that the growth rate decreased as irradiation power increases. The detailed irradiation power dependency analyzed using rate equation will be discussed later.

Figure 3.23a, b shows the TEM images of ZnO QDs grown for 3 days without and with light irradiation, respectively. The lattice fringes of ZnO QDs can be identified very clearly in the magnified images. The lattice spacings were estimated to be 0.274 nm (Fig. 3.23a) and 0.270 nm (Fig. 3.23b), which agree well with that of ZnO QDs along the c-axis [48], confirming sufficiently high-quality single-crystal growth. Figure 3.23c, d shows the distribution of the ZnO QD diameters measured by binarized TEM images shown in Fig. 3.23a, b, respectively. From these results we found that the average diameter grown with 325-nm laser irradiation was 4.1 nm, but it increased to 4.9 nm without light irradiation. These values agree well with the values estimated from the PL spectral peak wavelength in Fig. 3.22 (5.05 nm in diameter for ZnO QDs obtained through normal growth without light irradiation and 4.26 nm in diameter for ZnO QDs grown with 325-nm laser irradiation). We also found that the full width at half-maximum (FWHM) of the diameter distributions was 1.09 nm for ZnO QDs grown without light irradiation, whereas it was 0.71 nm for ZnO QDs grown with 325-nm light irradiation. This result indicates that the fluctuation in diameters decreased from 23% to 18% by introducing photo-desorption, confirming highly accurate particle size control.

The ZnO QD growth rate was analyzed using a rate equation:

$$dV/dt = \alpha S - \beta V I, \tag{3.2}$$

where V and S are the volume and the surface area of the QD, respectively. α and β are the proportionality constants. The first term of (3.2) represents the growth rate, which is proportional to the amount of the material to be adsorbed on the QD surface and is therefore proportional to S. The second term represents the desorption

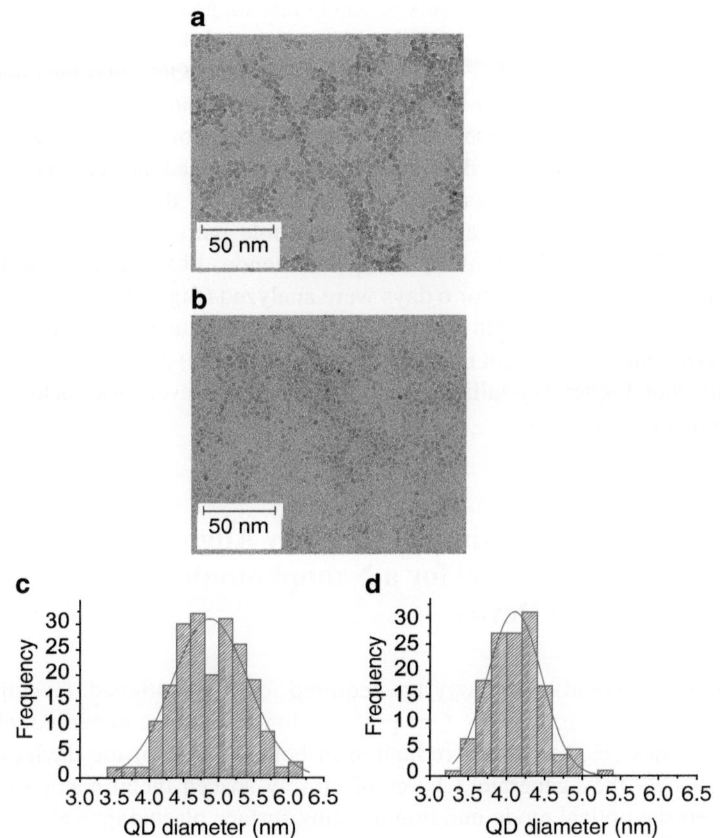

Fig. 3.23 (**a**) TEM image of ZnO QDs deposited without light irradiation. (**b**) TEM image of ZnO QDs deposited with 325-nm laser irradiation. (**c**) Size distribution of ZnO QDs produced without light irradiation. (**d**) Size distribution of ZnO QDs produced with 325-nm laser irradiation

rate, which is proportional to the light intensity absorbed by QDs and is therefore proportional to V. By assuming a spherical shape, S is equal to $4\pi(3V/4\pi)^{2/3}$.

By fitting (3.2) to the experimental values obtained without light irradiation ($I = 0$ mW/cm^2), the value of α was 2.81. Next, by fitting to the experimental values for $I = 8$ mW/cm^2, $\beta_{I=8\text{mw/cm}^2}$ was 0.032. Furthermore, $\beta_{I=4\text{mw/cm}^2}$ was 0.029 for $I = 4$ mW/cm^2, which agrees well with the result for $\beta_{I=8\text{mw/cm}^2}$. Thus, the above model is valid, indicating that particle sizes can be controlled in proportion to light intensity. Since we have confirmed that the spectral linewidth of ZnO single quantum structures remained constant at excitation power densities ranging from 0.5 W/cm^2 to 5 W/cm^2 of CW light source ($\lambda = 325$ nm) [49], indicating that the increase in the temperature is negligible. Therefore, it is possible to decrease the size variance below 10% by increasing the light intensity without adverse effect on QD properties.

The ability to excite defect levels of ZnO QDs suggests that the region causing these defect levels can be removed by photo-induced desorption. A defect level originates from an oxygen defect or an impurity in ZnO QDs, and the energy level corresponding to such a defect level is lower than the bandgap energy. Thus, as the amount of defect levels increases, the quantum efficiency of the ZnO QDs decreases. However, with this method, defect levels were removed preferentially because photo-induced desorption occurred in defect portions due to local oxidation–reduction reactions after the excited electron–hole pairs relaxed to those defect portions in ZnO QDs. To confirm this phenomenon, PL spectra, including long wavelengths, for QDs grown for 6 days were analyzed (Fig. 3.19b). Compared with the PL spectrum without light irradiation (black solid curve), the PL intensity of QDs grown with 325-nm light irradiation (red solid curve) decreased around 50 nm, indicating that higher crystallinity with fewer defect levels was achieved using photo-induced desorption.

3.6 Self-assembly Method of Linearly Aligning ZnO Quantum Dots for a Nanophotonic Signal Transmission Device

Innovations in optical technology are required for the continued development of information processing systems. One potential innovation, the increased integration of photonic devices, requires a reduction in both the size of the devices and the amount of heat they generate. Chains of closely spaced metal nanoparticles that can convert the optical mode into nonradiating surface plasmonic waves have been proposed as a way to meet these requirements [50,51]. However, one disadvantage is that they cannot break the plasmon diffraction limit. To overcome this difficulty, we developed nanophotonic signal transmission (NST) devices consisting of semiconductor quantum dots (QDs) (see Fig. 3.24) [3, 8, 52]. These NST devices operate using excitons generated in the QDs by optical near-field interactions between closely spaced QDs as the signal carrier. The exciton energy is transferred to adjacent QDs through resonant exciton energy levels, and therefore the optical beam spot may be decreased to the size of the QD. Moreover, NST devices using semiconductor QDs have higher transmission efficiency because QDs hardly couple with the lattice vibrational modes as opposed to transmission in metallic waveguides. This lattice vibration is the principal cause of large propagation losses in plasmonic waveguides. Here, we report a self-assembly method that aligns nanometer-sized QDs into a straight line along which photonic signals can be transmitted by optically near-field effects.

The NST device fabrication process requires the following:

(a) *Small size dispersion of QDs*: It is estimated that to fabricate an NST device consisting of 5-nm QDs and an efficiency equivalent of 97%, the size dispersion must be as small as ± 0.5 nm [8].

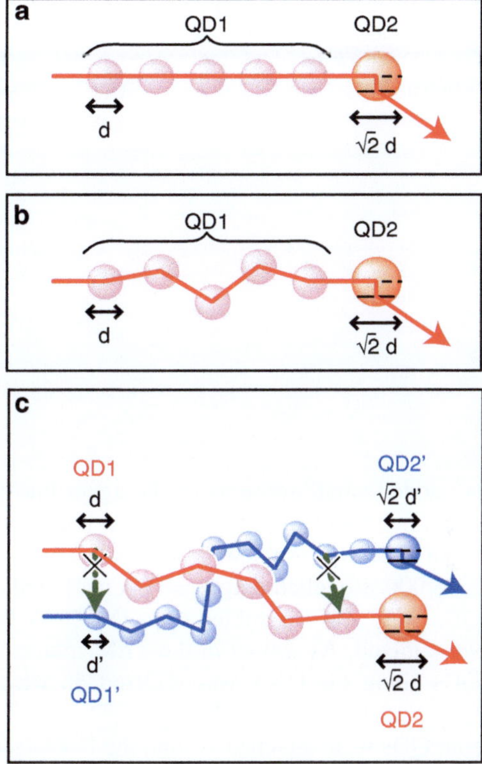

Fig. 3.24 Schematic of NST. Several function can be expected in NST, including (**a**) unidirectional energy flow is confirmed by the energy dissipation at the output QDs, (**b**) autonomous signal transmission, and (**c**) three-dimensional wavelength division multiplex

(b) *Subnanometer scale controllability in the separation between QDs*: Because the optical near-field coupling efficiency between adjacent QDs is determined by a Yukawa type function [53], the separation between QDs should be controlled to a subnanometer scale.

To meet these requirements, we developed a new technique for positioning and aligning QDs that yields precise separation. Figure 3.25 illustrates our approach to a self-assembling NST device with angstrom-scale spacing controllability among QDs using silane-based molecular spacers and deoxyribonucleic acid (DNA) [54, 55]. First, ZnO QDs 5 nm in diameter were synthesised using the sol–gel method [15, 56]. Typical transmission electron micrographic (TEM) images of synthesised ZnO dots are presented in Fig. 3.26a, b. The dark areas indicated by the dashed ellipses correspond to the ZnO QD. These images reveal lattice spacing matches for the c-plane (0.26 nm) and m-plane (0.28 nm) of wurtzite ZnO. These results confirmed that the fabricated ZnO QDs had high-quality single-crystal crystallinity. An average ZnO QDs diameter of 5.2 nm with a standard deviation (σ of 0.5 nm was determined from TEM images (Fig. 3.26c) and meets the first requirement (a).

Second, the surfaces of QD were coated with a silane coupling agent $N^+(CH_3)_3(CH_2)_3Si(OCH_3)_3$ with 0.6 nm in length. To avoid particle aggregation,

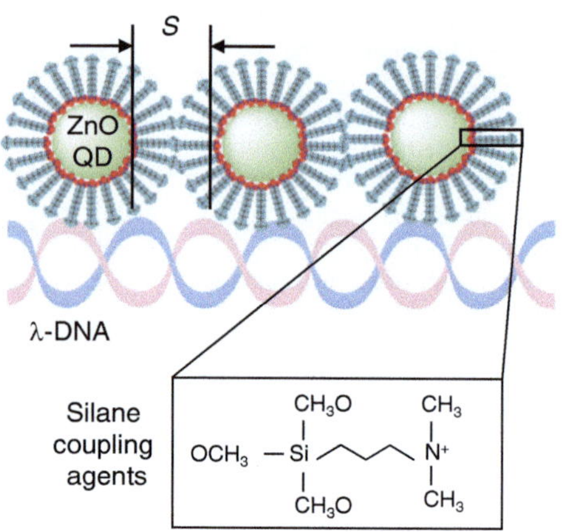

Fig. 3.25 Schematic of ZnO QD alignment along the λ DNA. S: separation between QDs

we added small amounts of the agent into the ZnO QD colloidal dispersion. The agent maintains the spacing between QDs and, because of its cationic nature, acts as an adhesive for the anionic DNA. Third, we used λ DNA (number of base pairs $=48,000$, stretched length $=16.4\ \mu$m) as the template to align the QDs so that the QDs were self-assembled onto the DNA by electrostatic interactions when they were mixed. As shown in the TEM image (Fig. 3.27a), dense packing of the ZnO QDs along the DNA was realized, in which the diameter of the DNA with the QDs was 15 nm. Considering a QD size of 5 nm and the DNA diameter of 2 nm, four QDs were attached around the DNA (see Fig. 3.27b, c). The 1.2-nm separation between QDs (S) was determined from the TEM image (Fig. 3.27d) and was in good agreement with twice the length of the silane coupling agents. Because the length of the silane coupling agents can be controlled by changing the number of CH_2 by 0.15 nm, this technique meets the second requirement (b). Despite the electrostatic repulsion between ZnO QDs, such high-density packing is due to the quaternary ammonium group of the silane coupling agent [54]. Quaternary ammonium groups and these QDs have highly condensed positive electrical changes on their particle surfaces. Because of these condensed charges, the QDs were densely adsorbed on the oppositely charged surfaces when their intervals were fixed by the stabilizer length.

To observe the optical properties of the NST, we stretched and straightened the QD-immobilised DNA on the silicon substrate using the molecular combing technique (Fig. 3.28) [57]. First, the silicon substrate was terminated with the silane coupling agent so that the anionic DNA was adsorbed on the cationic silicon substrate. Second, the solution including the DNA and the QDs was dropped onto the cationic silicon substrate. Finally, the glass substrate was slid over the droplet. To check the attachment of DNA–QDs on substrate, PL spectra were compared. For this purpose, four samples were prepared:

(a) DNA with QDs was dropped onto the cationic silicon substrate (see Fig. 3.29a).
(b) DNA with QDs was dropped onto the bare silicon substrate (see Fig. 3.29b).

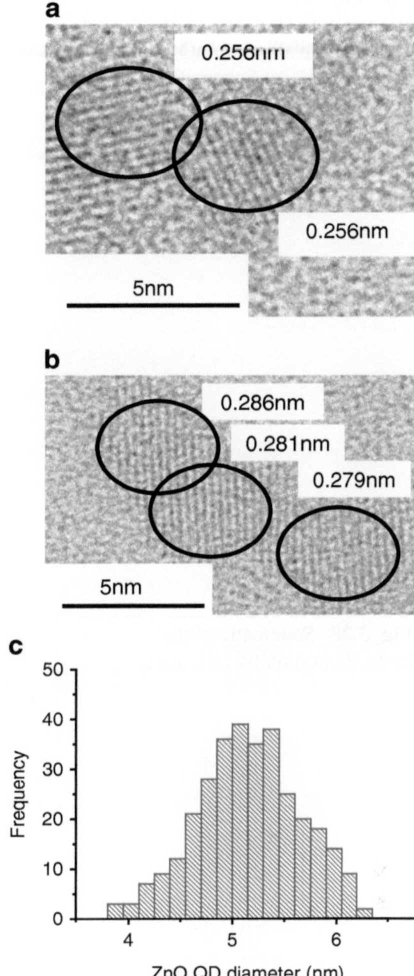

Fig. 3.26 TEM analysis of fabricated ZnO QDs. (**a**) and (**b**) Typical TEM pictures of the ZnO QDs. (**c**) ZnO QD diameter distribution

(c) QDs was dropped onto the cationic silicon substrate (see Fig. 3.29c).

(d) QDs was dropped onto the bare silicon substrate (see Fig. 3.29d).

Curves A–D in Fig. 3.29e, f show PL spectra of (a)–(d) in Fig. 3.29. These results confirmed that the anionic DNA is attached to the cationic silicon substrate. To confirm the alignment of DNA–QDs alignment, we obtained an emission image of the cyanine dye attached to the DNA using its 540-nm emission peak under halogen lamp illumination. As shown in the optical image taken with a charge-coupled device camera (Fig. 3.30), the DNA with QDs stretched in the direction determined by the slide direction of the glass substrate; also, these stretched DNA were found to be isolated.

Fig. 3.27 TEM analysis of aligned ZnO QDs. (**a**) TEM image of the aligned ZnO QDs. (**b**) and (**c**) Schematics of the QD alignment along the DNA. (**d**) Separation (S) distribution

Fig. 3.28 Schematic of the molecular combing technique

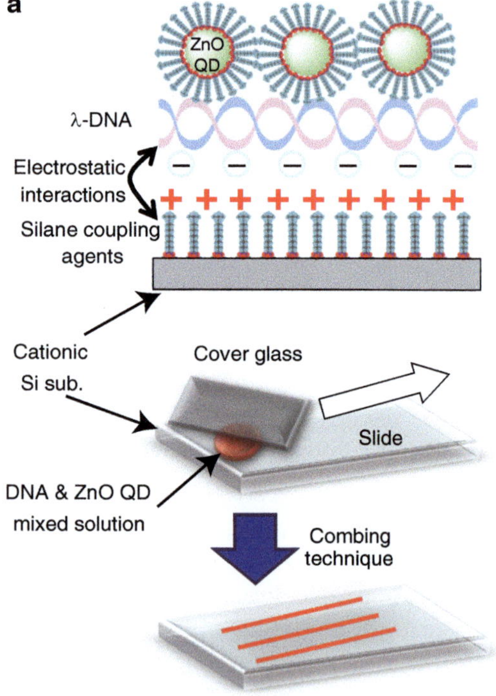

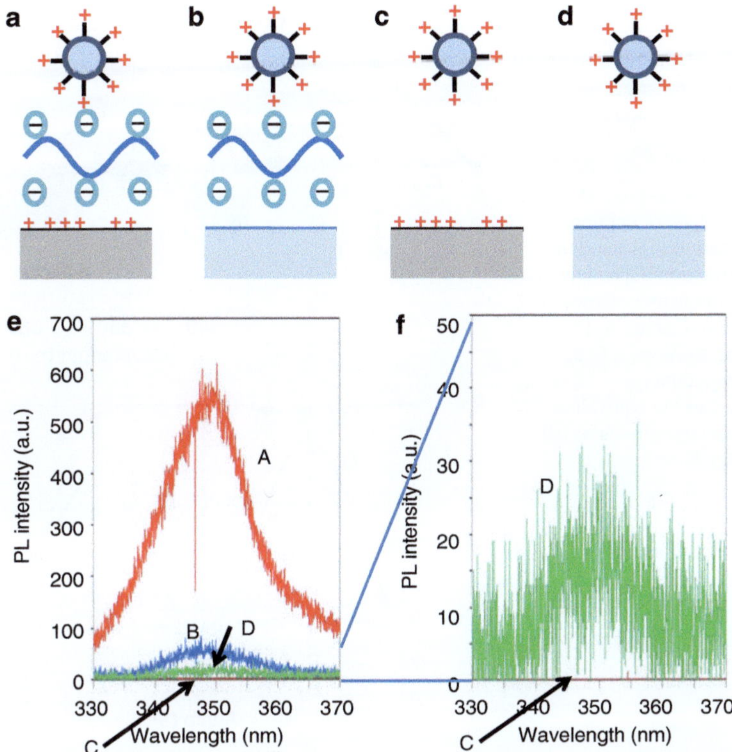

Fig. 3.29 Schematic of the samples preparation: (**a**) DNA with QDs was dropped onto the cationic silicon substrate, (**b**) DNA with QDs was dropped onto the bare silicon substrate, (**c**) QDs was dropped onto the cationic silicon substrate, and (**d**) QDs was dropped onto the bare silicon substrate. (**e**) Curves A–D show PL spectra of (**a**)–(**d**). (**f**) Magnified PL spectra of (**e**)

Fig. 3.30 (**a**) Charge-coupled device image of the stretched λ DNA. (**b**) Magnified image of (**a**)

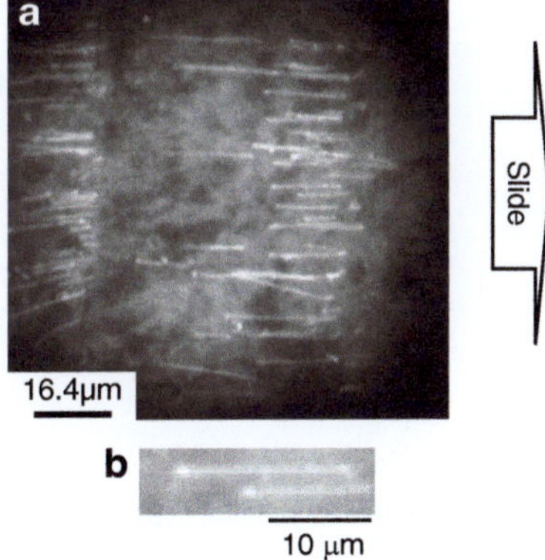

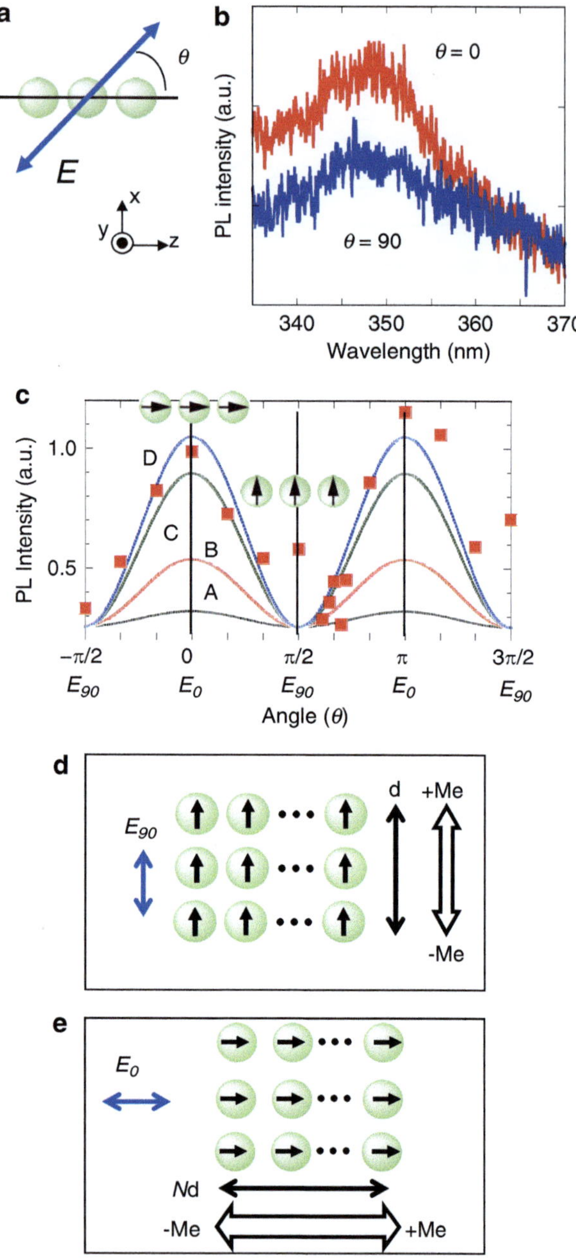

Fig. 3.31 Polarization dependence of the linearly aligned QD chain. (a) Incident light polarization dependence of PL intensity. θ: Polarization angle with respect to the direction along the QD chains (x-axis). (b) Typical PL spectra obtained at $\theta = 0$ and 90. (c) Incident light polarization dependence of the PL intensity obtained at $\lambda = 350$ nm. Curves A–D correspond to $N = 3, 6, 9$, and 10, respectively. Schematics of the equivalent total dipole strength under (d) E_{90} and (e) E_0

Using the linearly aligned ZnO QDs, we evaluated the photoluminescence (PL) polarization dependence. A 4th-harmonic of a Q-switched Nd:YAG laser ($\lambda = 266$ nm) with a spot size of approximately 2 mm was used to excite the ZnO QDs at various polarization angles (Fig. 3.31a). From the polarization dependence of the PL at a wavelength of 350 nm (Fig. 3.31b), corresponding to the ground state of 5-nm ZnO QDs, stronger PL emission was obtained by exciting the parallel polarization along the QD chains (E_0) than was obtained under the perpendicular polarization (E_{90}; Fig. 3.31c). Since the decay time of ZnO QDs is more than 20 times longer than the energy transfer time to adjacent QDs [56], it is possible that the dipoles between adjacent QDs were coupled by an optical near-field interaction, indicating that the signals were transmitted through the QD chain. Furthermore, QD chains have great dipolar strength (see the inset of Fig. 3.31c) that can be realized when the QDs are coherently coupled [58,59]. If M QDs are coherently coupled and the coherent length along the z-axis is N times greater than that along the x-axis, the equivalent total dipolar strength is given by $Me \times d$ (E_{90}; see Fig. 3.31d) and $MeNd$ ($E0$; see Fig. 3.31e), where e is the electrical charge excited in the QD and d is the coherent length along the x-axis, which is equivalent to the width of the QD chain. The resulting emission intensities are $(Me \times d)^2$ and $(Me \times Nd)^2$ for E_{90} and E_0, respectively. Therefore, we obtained N^2 times greater PL intensity with E_0 than with E_{90}. To evaluate the number of coherently coupled QDs, N, we fit the polarization intensity dependence PL(θ) using

$$\mathrm{PL}(\theta) = k(\sin^2 \theta + N^2\cos^2\theta) + \mathrm{const.}, \tag{3.3}$$

where k is a proportionality constant. As shown by the solid curves in Fig. 3.31c (curves A–D correspond to $N = 3, 6, 9$, and 10, respectively), the polarization dependence of the PL intensity was fitted using (3.3), and the value of N was estimated to be 10, indicating that the coherent length along the z-axis was 10-times greater than the width of the QD chain. Since the QD chain was 15 nm wide (see Fig. 3.27a), the coherent length along the QD chain was 150 nm. This value is 5 times larger than that in bulk ZnO [60]. This large coherent length along the QD chains originates from the reduction of phonon scattering in the QDs due to the decrease in the propagation length through ZnO for QDs.

References

1. T. Kawazoe, K. Kobayashi, J. Lim, Y. Narita, M. Ohtsu, Phys. Rev. Lett. **88**, 067404 (2002)
2. T. Yatsui, H. Jeong, M. Ohtsu, Appl. Phys. B **93**, 199 (2008)
3. M. Ohtsu, T. Kawazoe, T. Yatsui, M. Naruse, IEEE J. Sel. Top. Quantum Electron. **8**, 1404 (2008)
4. T. Yatsui, Y. Ryu, T. Morishima, W. Nomura, T. Kawazoe, T. Yonezawa, M. Washizu, H. Fujita, M. Ohtsu, Appl. Phys. Lett. **96**, 133106 (2010)
5. T. Kawazoe, K. Kobayashi, S. Sangu, M. Ohtsu, Appl. Phys. Lett. **82**, 2957 (2003)
6. T. Kawazoe, K. Kobayashi, K. Akahane, M. Naruse, N. Yamamoto, M. Ohtsu, Appl. Phys. B **84**, 243 (2006)
7. T. Kawazoe, K. Kobayashi, M. Ohtsu, Appl. Phys. Lett. **86**, 103102 (2005)
8. W. Nomura, T. Yatsui, T. Kawazoe, M. Ohtsu, J. Nanophoton. **1**, 011591 (2007)
9. T. Kawazoe, K. Kobayashi, S. Sangu, M. Ohtsu, J. Microsc. **209**, 261 (2003)
10. K. Nishibayashi, T. Kawazoe, M. Ohtsu, K. Akahane, N. Yamamoto, J. Luminescence **129**, 1912 (2009)

11. T. Matsumoto, M. Ohtsu, K. Matsuda, T. Saiki, H. Saito, K. Nishi, Appl. Phys. Lett. **75**, 3346 (1999)
12. H.D. Sun, T. Makino, Y. Segawa, M. Kawasaki, A. Ohtomo, K. Tamura, H. Koinuma, J. Appl. Phys. **91**, 1993 (2002)
13. S. Besner, A.V. Kabashin, F.M. Winnik, M. Meunier, Appl. Phys. A **93**, 955 (2008)
14. H.Z. Wu, D.J. Qiu, Y.J. Cai, X.L. Xu, N.B. Chen, J. Crystal Growth **245**, 59 (2002)
15. E.A. Meulenkamp, J. Phys. Chem. B **102**, 5566 (1998)
16. E.M. Hendriks, Z. Phys. B **57**, 307 (1984)
17. T. Torimoto, S. Murakami, M. Sakuraoka, K. Iwasaki, K. Okazaki, T. Shibayama, B. Ohtani, J. Phys. Chem. B **110**, 13314 (2006)
18. H. Koyama, N. Koshida, J. Appl. Phys. **74**, 6365 (1993)
19. M. Brust, C.J. Kiely, Colloids Surf. A, **202**, 175 (2002)
20. A.P. Alivisatos, Science **271**, 933 (1996)
21. P. Yang, Nature **425**, 243 (2003)
22. M.D. Austin, H. Ge, W. Wu, M. Li, Z. Yu, D. Wasserman, S.A. Lyon, S.Y. Chou, Appl. Phys. Lett. **84**, 5299 (2004)
23. G.M. Whitesides, B. Grzybowski, Science **295**, 2418 (2002)
24. Y. Yin, Y. Lu, Y. Xia, J. Mater. Chem. **11**, 987 (2001)
25. Y. Cui, M.T. Björk, J.A. Liddle, C. Sönnichsen, B. Boussert, A.P. Alivisatos, Nano Lett. **4**, 1093 (2004)
26. G. Frens, Nature Phys. Sci. **241**, 20 (1973)
27. T. Yatsui, M. Kourogi, M. Ohtsu, Appl. Phys. Lett. **79**, 4583 (2001)
28. N.D. Denkov, O.D. Velev, P.A. Kralchevsky, I.B. Ivanov, H. Yoshimura, L. Nagayama, Nature **361**, 26 (1993)
29. T. Yatsui, S. Takubo, J. Lim, W. Nomura, M. Kourogi, M. Ohtsu, Appl. Phys. Lett. **83**, 1716 (2003)
30. A. Wokaun, J.P. Gordon, P.F. Liao, Phys. Rev. Lett. **48**, 957 (1982)
31. G.T. Boyd, Th. Rasing, J.R.R. Leite, Y.R. Shen, Phys. Rev. B **30**, 519 (1984)
32. J. Bosbach, D. Martin, F. Stietz, T. Wenzel, F. Träger, Appl. Phys. Lett. **74**, 2605 (1999)
33. K.F. MacDonald, V.A. Fedotov, S. Pochon, K.J. Ross, G.C. Stevens, N.I. Zheludev, W.S. Brocklesby, V.I. Emel'yanov, Appl. Phys. Lett. **80**, 1643 (2002)
34. A. Sommerfeld *Optics* (Academic Press, New York, 1954)
35. M. Quinten, A. Leitner, J.R. Krenn, F.R. Aussenegg, Opt. Lett. **23**, 1331 (1998)
36. W. Nomura, T. Yatsui, M. Ohtsu, Appl. Phys. Lett. **86**, 181108 (2005)
37. W. Nomura, T. Yatsui, M. Ohtsu, Appl. Phys. B **84**, 257 (2006)
38. T. Ito, M. Ogino, T. Yamada, Y. Inao, T. Yamaguchi, T. Mizutani, R. Kuroda, J. Photopoly. Sci. Tech. **18**, 435 (2003)
39. S.Y. Chou, P.R. Krauss, W. Zhang, L. Guo, L. Zhuang, J. Vac. Sci. Technol. B **15**, 2897 (1997)
40. M.D. Austin, H. Ge, W. Wu, M. Li, Z. Yu, D. Wasserman, S.A. Lyon, S.Y. Chou, Appl. Phys. Lett. **84**, 5299 (2004)
41. The computer simulations in this paper are performed by a FDTD-based program, Poynting for Optics, a product of Fujitsu, Japan
42. E.D. Palik *Handbook of Optical Constants of Solids* (Academic, New York, 1985)
43. H. Raether *Surface Plasmons* (Springer-Verlag, Berlin, 1988)
44. J. Haisma, M. Verheijen, K. van den Heuvel, J. van den Berg, J. Vac. Sci. Technol. B **14**, 4124 (1996)
45. L. Zhang, L.W. Yin, C. Wang, N. Lun, Y. Qi, Appl. Mater. Inter. **2**, 1769 (2010)
46. A.D. Yoffe, Adv. Phys. **51**, 799 (2002)
47. A. Wood, M. Giersig, M. Hilgendorff, A. Vilas-Campos, L.M. Liz-Marzan, P. Mulvaney, Aust. J. Chem. **56**, 1051 (2003)
48. U. Ozfu, L. Alivov Ya, C. Liu, A. Teke, M.A. Reshchikov, S. Dogan, V. Avrutin, S.J. Cho, H. Horkoc, J. Appl. Phys. **98**, 041301 (2005)
49. T. Yatsui, M. Ohtsu, J. Yoo, S.J. An, G.-C. Yi, Appl. Phys. Lett. **87**, 033101 (2005)

50. S.A. Maier, P.G. Kik, H.A. Atwater, S. Meltzer, E. Harel, B.E. Koel, A.A.G. Requicha, Nature Mater. **2**, 229 (2003)
51. W. Nomura, M. Ohtsu, T. Yatsui, Appl. Phys. Lett. **86**, 181108 (2005)
52. C.-J.Wang, L. Huang, B.A. Parviz, L.Y. Lin, Nano Lett. **6**, 2549 (2006)
53. M. Ohtsu, K. Kobayashi, T. Kawazoe, S. Sangu, T. Yatsui, IEEE J. Sel. Top. Quantum Electron. **8**, 839 (2002)
54. T. Yonezawa, S. Onoue, N. Kimizuka, Chem. Lett. **31**, 1172 (2002)
55. M.G. Warner, J.E. Hutchison, Nature Mater. **2**, 272 (2003)
56. T. Yatsui, H. Jeong, M. Ohtsu, Appl. Phys. B: Lasers Opt. **93**, 199 (2008)
57. H. Oana, M. Ueda, M. Washizu, Biochem. Biophys. Res. Commun. **265**, 140 (1999)
58. M. Tammer, L. Horsburgh, A.P. Monkman, W. Brown, H. Burrows, Adv. Funct. Mater. **12**, 447 (2002)
59. M. Campoy-Quiles, Y. Ishii, H. Sakai, H. Murata, Appl. Phys. Lett. **92**, 213305 (2008)
60. B. Gil, A.V. Kavokin, Appl. Phys. Lett. **81**, 748 (2002)

Chapter 4
Phonon-Assisted Process

4.1 Dressed-Photon and Phonon

The optical near field is a virtual photon that can couple with an excited electron. In the coupled state, it is known as a dressed photon and is a quasiparticle (Fig. 4.1). The energy of the dressed photon, $h\nu_{dp}$, is larger than the energy of a free photon, $h\nu$, due to coupling with and excited electron.

Because nanomaterials consist of electrons and a crystal lattice, the dressed photon can couple with multimode of phonons. As a result, the dressed photon can dress the energy of phonons in a coherent state. The creation operator of the dressed photon and phonon is described as

$$\hat{\alpha}_i^\dagger = \tilde{a}_i^\dagger \left\{ -\sum_{p=1}^{N} \frac{\chi_{i,p}}{\Omega_p} \exp\left(\hat{b}_p^\dagger - \hat{b}_p\right) \right\}, \tag{4.1}$$

where, $\tilde{a}_i^\dagger$ is the creation operator of the dressed photon localized at the ith site in the crystal lattice, N the total number of sites, χ_{ip} the coupling constant between the dressed photon at the ith site and the phonon of mode p, and Ω_p the natural angular frequency of the phonon of mode p. The creation and annihilation operators of the phonon are $(b_p{}^\dagger, b_p)$ [1–4]. The coupled state of the dressed photon and the coherent phonon (dressed photon and phonon) is a quasiparticle. The dressed photon and phonon can be generated when the size of the material is small enough that the lattice vibration is excited coherently. When the lattice vibration is excited incoherently, the energy dissipates, heating the particles.

The energy of the dressed photon and phonon, E_{DP-P}, is larger than that of the dressed photon, E_{DP}, and the incident free photon, E_{FP}:

$$E_{FP} < E_{DP} < E_{DP-P}. \tag{4.2}$$

T. Yatsui, *Nanophotonic Fabrication*, Nano-Optics and Nanophotonics,
DOI 10.1007/978-3-642-24172-7_4, © Springer-Verlag Berlin Heidelberg 2012

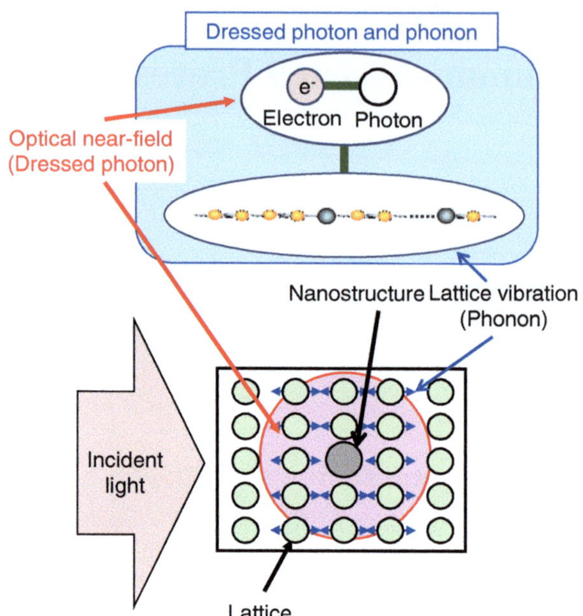

Fig. 4.1 Schematic of the dressed photon and phonon

The increase in energy originates from the contributions of the electron and phonon. Use of the dressed photon and phonon has facilitated novel technology on nanoscale.

4.2 Angstrom Scale Flattening Process

An ultra-flat substrate (sub-nanometer scale roughness) is required for the manufacture of high-quality, extreme UV (EUV) optical components, high-power lasers, and ultrashort-pulse lasers, plus future photonic devices at the sub-100-nm scale. It is estimated that the required surface roughness, R_a, will be less than $1\,\text{Å}$ [5]. This R_a value is an arithmetic average of the absolute values of the surface height deviations measured from a best-fit plane, and is given by

$$R_a = \frac{1}{l} \int_0^l |f(x)|\mathrm{d}x$$

$$\cong \frac{1}{n} \sum_{i=1}^{n} |f(x_i)|, \tag{4.3}$$

where $|f(x_i)|$ are absolute values measured from the best-fit plane and l is the evaluation length (see Fig. 4.2). Physically, $\mathrm{d}x$ corresponds to the spatial resolution of the measurement of $f(x)$, and n is the number of pixels in the measurement

Fig. 4.2 Schematic of R_a

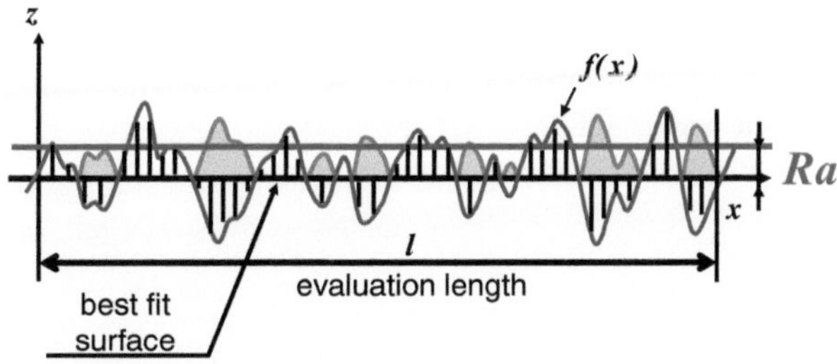

($n = l/\mathrm{d}x$). Conventionally, chemical–mechanical polishing (CMP) is used to achieve flat surfaces [6]. However, with CMP, it is difficult to reduce R_a to less than 2 Å, as the polishing pad roughness is typically around $10\,\mu$m and the diameters of the polishing particles in the slurry are as large as 100 nm. In addition, polishing causes scratches or digs due to the contact between the polishing particles and/or the impurities in the slurry and the substrate.

Our interest in applying an optical near field to nanostructure fabrication was generated because of its high-resolution capability—beyond the diffraction limit— and because of its novel photochemical properties, whereby the reaction is classified as nonadiabatic due to its energy transfer via a virtual exciton-phonon-polariton [3,7]. In this chemical vapor deposition, photodissociation of the molecules is driven by the light source at a lower photon energy than the molecular absorption edge by a multiple-step excitation via vibrational energy levels [8]. Following this process, we propose a novel method of polishing using phonon-assisted optical near-field etching.

4.2.1 *Phonon-Assisted Optical Near-Field Etching*

A continuous wave laser ($\lambda = 532$ nm) was used to dissociate the Cl_2 gas through a phonon-assisted photochemical reaction. The photon energy is smaller than that corresponding to the absorption edge of Cl_2 ($\lambda = 400$ nm) [9], so the Cl_2 adiabatic photochemical reaction is avoided. However, because the substrate has nanometer-scale surface roughness, the generation of a strong optical near field on the surface is expected from simple illumination, with no focusing required (Fig. 4.3a). Since a virtual exciton-phonon-polariton can be excited on this roughness, a higher molecular vibrational state can be excited than on the flat part of the surface, where there is no virtual exciton-phonon-polariton. Cl_2 is therefore selectively photodissociated wherever the optical near field is generated. These dissociated Cl_2 molecules then etch away the surface roughness, and the etching process automatically stops when the surface becomes flat (Fig. 4.3b).

We used 30-mm-diameter planar synthetic silica substrates built by vapor-phase axial deposition with an OH group concentration of less than 1 ppm [10]. The substrates were preliminarily polished by CMP prior to the phonon-assisted

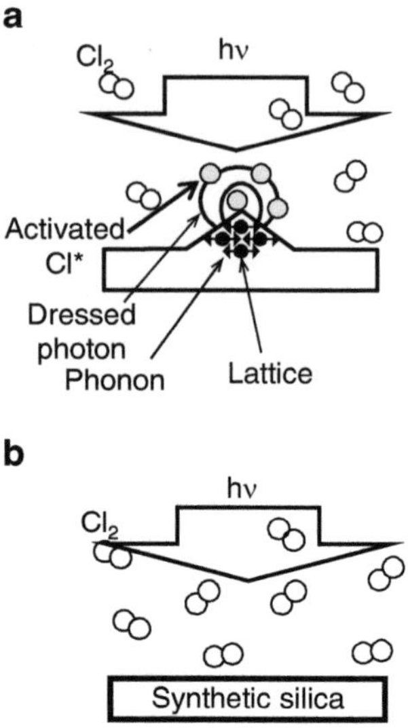

Fig. 4.3 Schematic of the near-field etching (**a**) during the etching process and (**b**) after etching

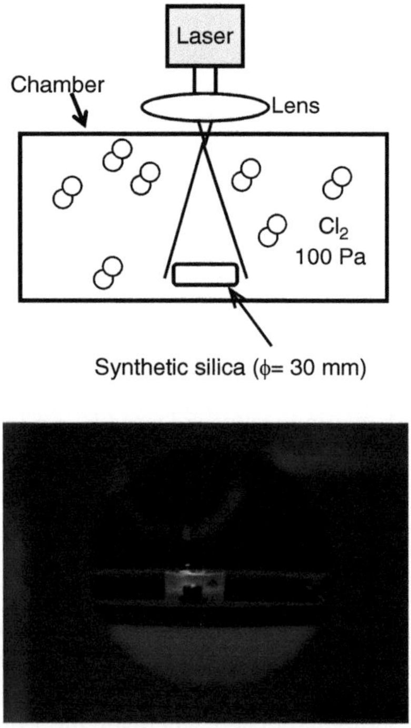

Fig. 4.4 Schematic of the experimental setup

optical near-field etching, which was performed at a Cl$_2$ pressure of 100 Pa at room temperature with a continuous wave laser ($\lambda = 532$ nm) having a uniform power density of 0.28 W/cm^2 (see Fig. 4.4). Surface roughness was evaluated using

Fig. 4.5 Schematic of the AFM measurement

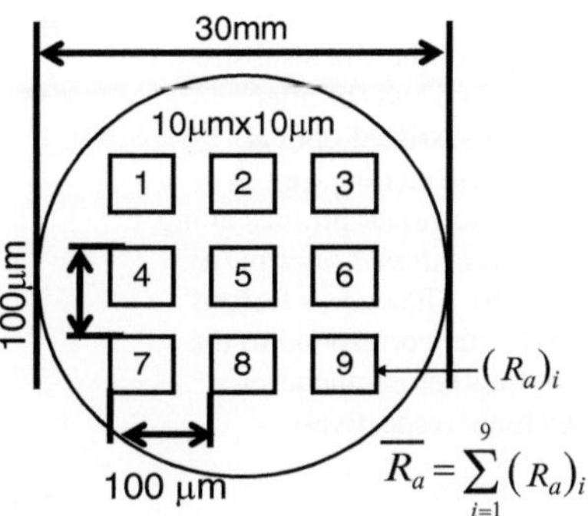

an atomic force microscope (AFM). Since the scanning area of the AFM was much smaller than the substrate, we measured the surface roughness, R_a, in nine representative areas, each $10\,\mu$m $\times$ $10\,\mu$m, separated by $100\,\mu$m (see Fig. 4.5). The scanned area was 256×256 pixels with a spatial resolution of $40\,$nm. The average value, $\bar{R}_a$, of the nine R_a values obtained before the phonon-assisted optical near-field etching and evaluated through the AFM images, was $2.36 \pm 0.02\,$Å. We cleaned the substrate ultrasonically using deionized water and methanol before and after the phonon-assisted optical near-field etching.

Figure 4.6a, b shows typical AFM images of the silica substrate area before and after phonon-assisted optical near-field etching, respectively. Note that the surface roughness was drastically reduced, as supported by the cross-sectional profiles along the dashed white lines in Fig. 4.6a, b (see Fig. 4.6c). We found a dramatic decrease in the value of the peak-to-valley roughness from $1.2\,$nm (dashed curve) to $0.5\,$nm (solid curve). Furthermore, note that the scratch seen in the AFM image before phonon-assisted optical near-field etching has disappeared. This indicates that rougher areas of the substrate had a higher etching rate, possibly because of greater intensity of the optical near field, leading to a uniformly flat surface over a wide area.

Figure 4.7a shows the etching time dependence of $\bar{R}_a$. We found that $\bar{R}_a$ decreases as the etching time increases. The minimum $\bar{R}_a$ was $1.37\,$Å at an etching time of $120\,$min, while the minimum R_a among the nine areas was $1.17\,$Å. Because the process is performed in a sealed chamber, the saturation in the decrease of $\bar{R}_a$ might originate from the decrease in the Cl_2 partial pressure during etching. A further decrease in $\bar{R}_a$ would be expected under constant Cl_2 pressure. Figure 4.7b shows the time dependence of the standard deviation of R_a (ΔR_a), which was obtained in one scanning area. We found a dramatic decrease in ΔR_a after $60\,$min, although we also found an increase in ΔR_a in the early stages of phonon-assisted optical near-field etching. This might have been caused by impurities, such as OH, on the substrate surface.

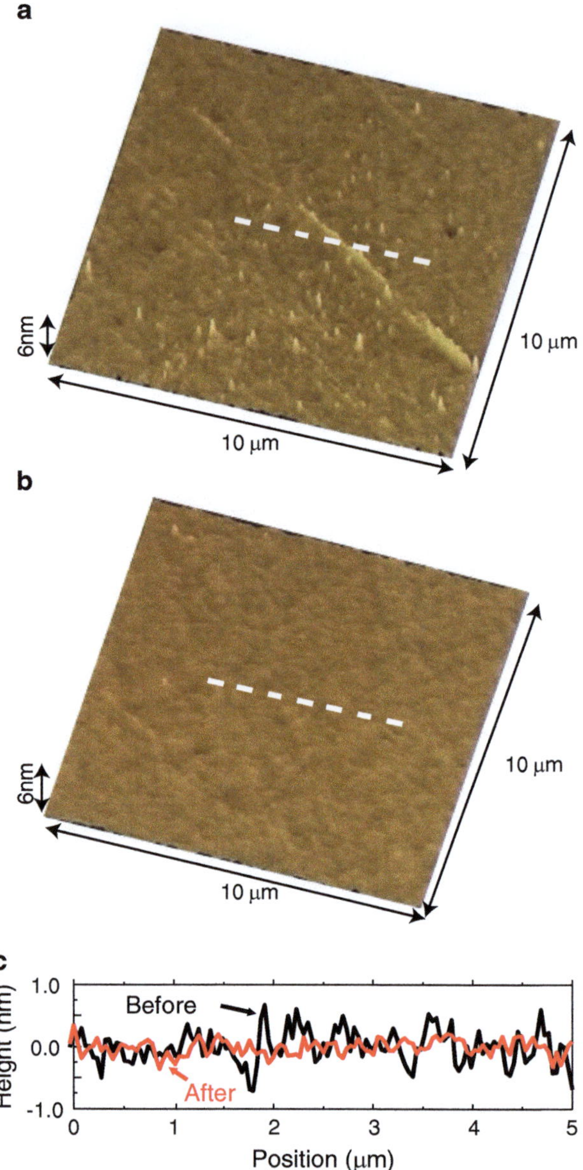

Fig. 4.6 Typical AFM images of the silica substrate (**a**) before and (**b**) after phonon-assisted optical near-field etching. (**c**) Cross-sectional profiles along the *white dotted lines* in (**a**) and (**b**). The *cueves* Before and After correspond to the profiles before and after etching, respectively

4.2.2 *In situ Real-Time Monitoring of Changes in the Surface Roughness During Phonon-Assisted Optical Near-Field Etching*

In a previous near-field etching study [11], the reduction in R_a was confirmed using an atomic force microscope (AFM) after etching. Because the scanning area of

Fig. 4.7 The etching time dependence of (**a**) the average R_a ($\bar{R}_a$) and (**b**) the standard deviation of R_a (ΔR_a)

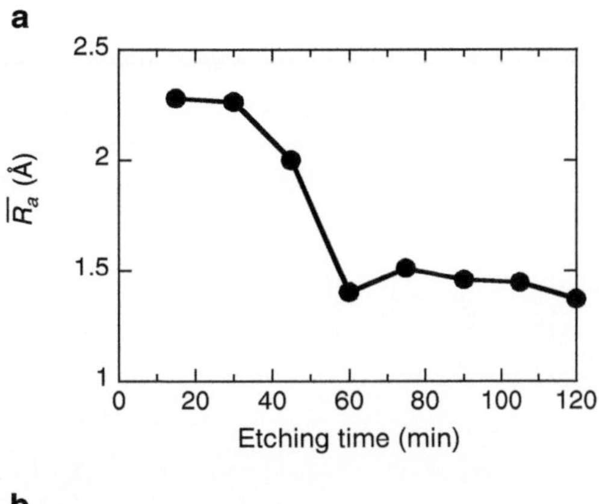

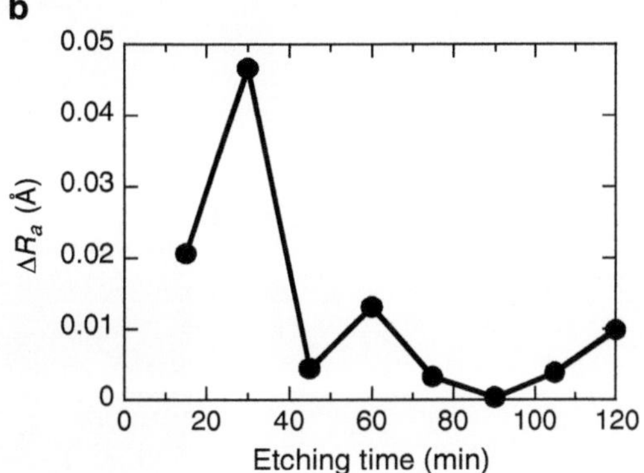

the AFM measurement was restricted to less than 10 μm $\times$ 10 μm, R_a had to be obtained as an average of several scanning areas. Furthermore, scattering loss reduction at the surface could not be confirmed.

For the present investigations, we performed an in situ real-time monitoring of the scattered light at the substrate surface during near-field etching. In addition, we obtained the correlation length of the surface roughness, allowing estimation of the structural change due to the near-field etching. These approaches allowed us to optimize the etching parameters for the further decrease surface roughness.

We used 30-mm-diameter planar synthetic silica substrates synthesized by vapor-phase axial deposition with an OH group concentration of less than 1 ppm [9]. The substrates were preliminarily polished by CMP prior to the near-field etching. The near-field etching was done with a CW laser ($\lambda = 532$ nm) having a uniform power density of 0.28 W/cm^2 over the substrate (see Fig. 4.8). The Cl$_2$ pressure in the chamber was maintained at 100 Pa at room temperature. The scattered light was generated with a CW laser with wavelength 632 nm power density 0.127 W/cm^2 and spot size diameter 1 mm. As this photon energy is lower than both the absorption band edge energy of Cl$_2$ ($\lambda = 400$ nm) and the light source for the near-field etching ($\lambda = 532$ nm), it does not contribute to the near-field etching. To increase the sensitivity and selectivity for detection of the scattered light, the incident light

Fig. 4.8 (**a**) Schematic of the in situ real time monitoring of the surface roughness during the near-field etching. (**b**) Photograph of the set-up

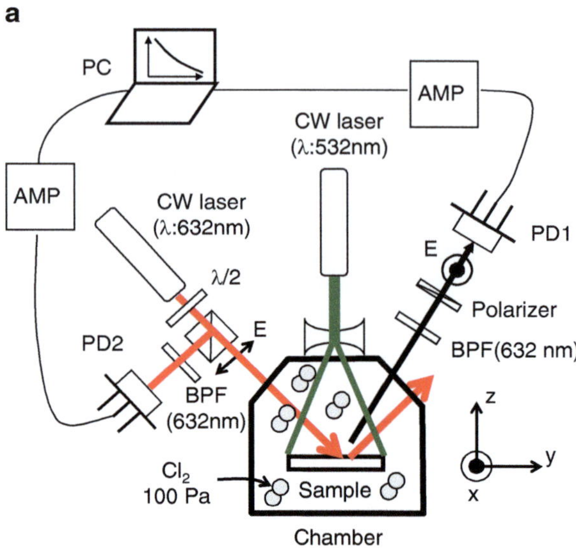

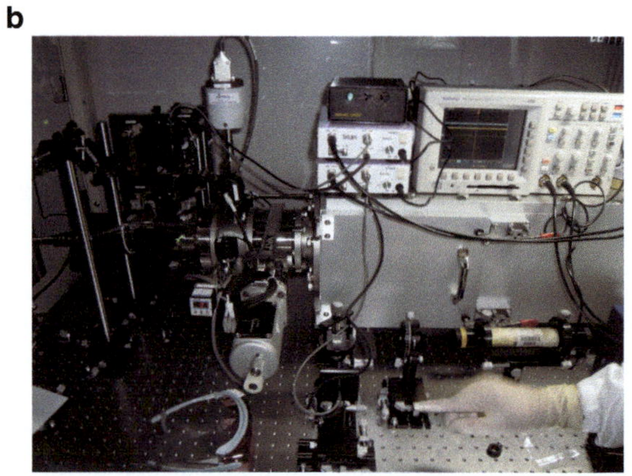

was polarized perpendicular to the x-axis (P-polarized light), and the polarized light parallel to the x-axis (S-polarized light) was detected after passing through a $\lambda = 632\,\mathrm{nm} \pm 1.5\,\mathrm{nm}$ band pass filter (BPF) and a polarizer. The scattered light intensity, detected by a photodetector (PD1), was normalized to the incident light intensity detected by another photodetector (PD2). The surface roughness was also evaluated with the AFM after near-field etching. As the scanning area of the AFM was much smaller than the substrate, we measured the surface roughness R_a in nine representative areas, each $10\,\mu\mathrm{m} \pm 10\,\mu\mathrm{m}$ and separated by $100\,\mu\mathrm{m}$. The scanned area was 8,192 (x-axis) × 256 (y-axis) pixels with a spatial resolution of 1.2 nm (x-axis) and 40 nm (y-axis), respectively.

Figure 4.9a, b shows typical AFM images of the scanned $10 \times 10\mu\mathrm{m}$ silica substrate area before and after 30 min of near-field etching, respectively. Figure 4.9c

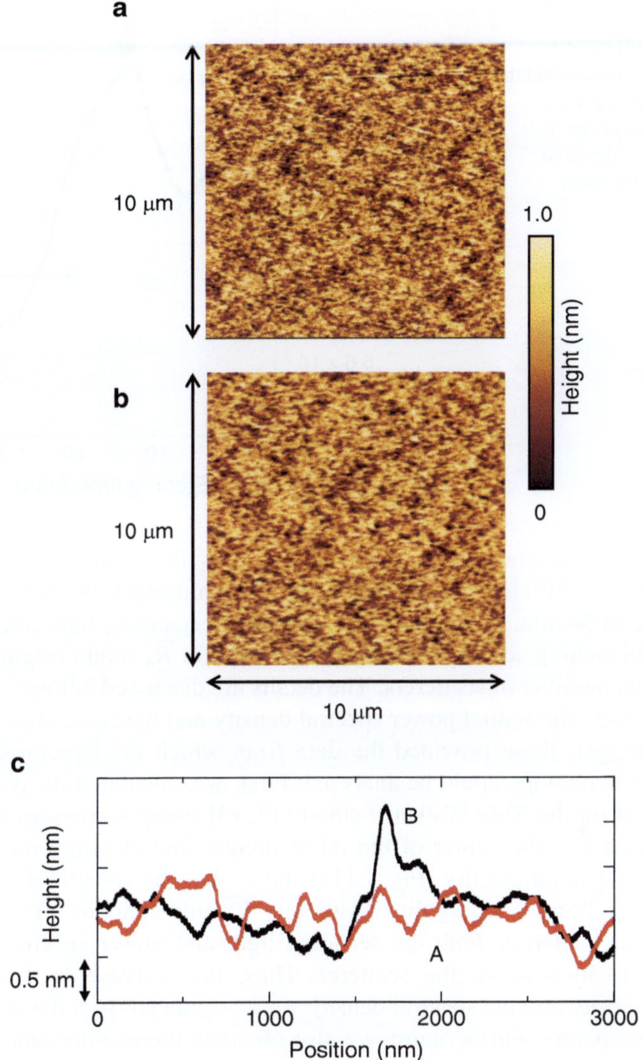

Fig. 4.9 Schematic of the in-situ real-time monitoring of the surface roughness during the near-field etching

shows the cross-sectional profiles of Fig. 4.9a, b, where the near-field etching yielded a dramatic decrease in peak-to-valley measurements from 1.73 nm (curve B) to 1.05 nm (curve A).

The solid curve in Fig. 4.10 shows the normalized scattered-light intensity, obtained during the near-field etching. The normalized scattered-light intensity was at maximum at an etching time of 13 min (indicated by the downward arrow in Fig. 4.10), and it decreased steadily as etching time increased. The etching time

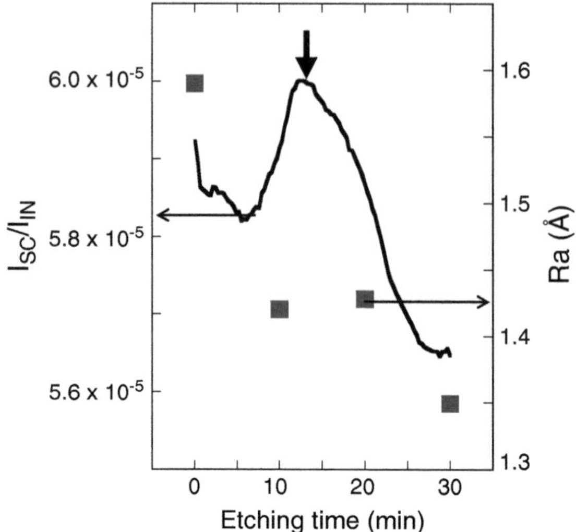

Fig. 4.10 The etching time dependence of the normalized scattered light intensity (I_{SC}/I_{IN}, *solid curve*) and the $\overline{R_a}$ (*solid squares*). I_{SC}, detected scattered intensity using PD1; I_{IN}, detected incident light intensity using PD2

dependence of the average $\overline{R_a}$ for the nine $R_a s$, also plotted in this figure, indicates $\overline{R_a}$ decreases as the etching time increases. The minimum $\overline{R_a}$ was 1.35 Å at an etching time of 30 min. This increase in normalized scattered light intensity during the near-field etching accompanied by the decrease in $\overline{R_a}$ might originate from the increase in the number of scatterers. The details are discussed below.

We calculated the spatial power spectral density and the correlation length from the AFM images; these provided the data from which the structural changes in the surface morphology could be analyzed. First, we calculated the power spectral density based on the Yule-Walker method [12, 13] using a cross-sectional profile along the x-axis at the center of the AFM image. The etching time dependence of the power spectral density (Fig. 4.11a) shows that the spectral density reached maximum at a 20-min etching time. Although $\overline{R_a}$ only contains information about the individual scatterers, both the scattered light and power spectra yield lateral surface information about the scatterer. Thus, the increase in the normalized scattered intensity and the spectral density were originated from the increase in the number of scatterers. Furthermore, we also obtained the etching time dependence of the lower- and higher-spatial frequency components of the spectral density. Figure 4.11b shows the results, where the low-and high-frequency components were fixed at scales of 1.0×10^{-3} (1/nm) (solid circles) and 1.3×10^{-1} (1/nm) (solid squares), respectively. From the lines fitted to these values by the least-square method, it was found that the low-frequency component decreased with increased etching time, whereas the high-frequency components increased. These results indicate that the size of the scatterer decreased as etching time increased. To confirm this postulate, we calculated the correlation length of the AFM images.

For the given surface function $f(u)$, the correlation function, $C_f(u)$, defined by

$$C_f(u) = \lim_{L \to \infty} \frac{1}{2L} \int_{-L}^{+L} f(z) f(z+u) \, du \qquad (4.4)$$

Fig. 4.11 (**a**) Etching time dependence of the power spectrum. (**b**) Etching time dependence of the spectral density at a scale of 1.0×10^{-3} (1/nm) (SD_{Low}) and spectral density at high frequency of 1.3×10^{-1} (1/nm) (SD_{High})

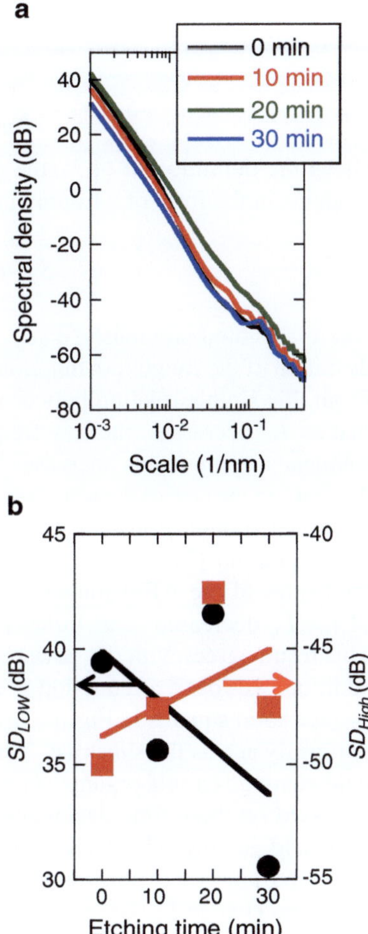

is a measure of the correlation between two points on the surface separated by a distance u. For stationary stochastic processes, the correlation function is a monotonically decreasing function of the correlation length L_c, where L_c corresponds to the average protrusion width in the surface profile, i.e., the width of the scatterer. This correlation function is well approximated with an exponential model of the form [14]:

$$C_f(u) = \delta_f{}^2 \exp(-u/L_c), \tag{4.5}$$

where δ_f is the standard deviation of the roughness function. Another important roughness quantity is its power spectrum, $S(\sigma)$, where σ represents the spatial frequencies present in the roughness function. The correlation function $C_f(u)$ and the power spectrum $S(\sigma)$ are linked to each other through the Wiener-Khintchine relations [15]

$$C_f(u) = \int_{-\infty}^{\infty} \exp\left(i\sigma u\right) S_f(\sigma) d\sigma \tag{4.6}$$

$$S_f(\sigma) = \frac{1}{2\pi} \int_{-\infty}^{\infty} \exp\left(-i\sigma u\right) C_f(\sigma) du \tag{4.7}$$

Using the definition of (4.7), the power spectrum associated with the exponential model is in the form of a Lorentzian

$$S_f(\sigma) = \frac{\delta_f^2}{\pi} \frac{L_c}{1 + L_c \sigma^2}, \tag{4.8}$$

where, as noted previously, σ, δ_f, and L_c are the spatial frequencies, the standard deviation of the roughness function, and the correlation length, respectively [14, 15]. From this simple relation between power spectrum and correlation length, we see that as L_c decreases, the low-frequency components decrease, whereas the high-frequency components increase. This can be understood by considering that a decrease in the size of the scatterer results in a decrease in the low-frequency components and simultaneously results in an increase in the high-frequency components. Using (1) and (2), we derived L_c and δ_f of surface profile along the x-axis at the center of the AFM images. Figure 4.11a shows the etching time dependence of L_c; L_c decreased to as little as 80 nm as the near-field etching time proceeded. This result agrees with the power spectral density change shown in Fig. 4.11a, b. In addition, we found a reduction in the standard deviation of the roughness function δ_f (see solid squares in Fig. 4.12a), indicating that the substrate surface was etched uniformly across the substrate. This high spatial uniformity indicated that the near-field etching is a self-organized process.

Based on these time dependences, we then considered structural change during near-field etching. The decrease in L_c rules out the possibility of structural change when the number of the scatterers remains constant (Fig. 4.12b) because for this structural change L_c, i.e., the width of the scatterer, remains constant. The increase in scattered light intensity and spectral density should originate from the increase in the number of scatterers, as shown schematically by the solid line in Fig. 4.12c. This structural change is supported by the fact that the L_c halved at an etching time of 10 min (Fig. 4.12a).

To visually understand the above postulate, we analyzed the AFM images numerically. First, the AFM images were digitized with a threshold height of 0.5 nm. Figure 4.13a–d shows the respective AFM images for 0–30 min etching time; the white area corresponds to the protruding area, i.e., the scatterer. Second, we evaluated the diameters of the scatterer by approximating them as circles of equal area. Figure 4.13e shows the diameter distributions of the scatterers fitted by the log-normal curves. Here, we found that the number of scatterers increased with etching times of 10 min and 20 min, whereas the peak diameter decreased. Figure 4.13f shows the comparison of L_c, previously presented as the solid circles in Fig. 4.12a, and the obtained peak diameters of the scatterers. The coincidence of both etching time dependences agrees with the above postulate that the width of the

Fig. 4.12 (a) Etching time dependence of the correlation length L_c and the standard deviation of the roughness function δ_f. (b) and (c) Schematic of structural change of the scatterer. *Dashed, solid, and dash-dotted lines* show the surface profiles before, during, and after near-field etching, respectively

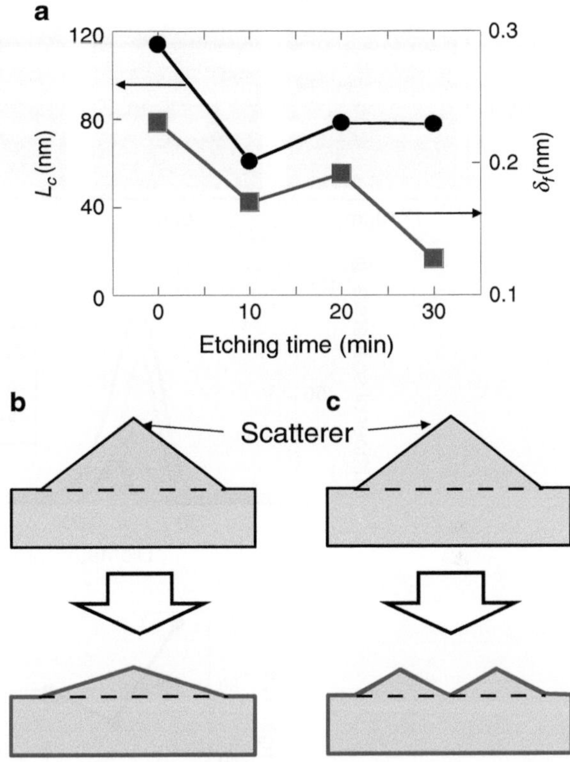

scatterers decreases as etching time increases. Figure 4.13g shows the comparison of the number of the scatterers and the scattered light intensity (I_{SC}/I_{IN}), previously plotted as the solid line in Fig. 4.10. This comparison indicates that both values are maximum at an etching time of 10 min, meaning that the near-field etching produces smaller scatterers, i.e., an increase in the number of scatterers, resulting in an increase in the scattered light intensity.

4.2.3   Self-organized Near-Field Etching of the Sidewalls of Glass Corrugations

In addition to possessing this unique organizing property, near-field etching is a non-contact method (thus eliminating the need for polishing pads), with anticipated applications to various three-dimensional surfaces, including concave and convex lenses, diffraction gratings, and the inner wall surfaces of a cylinder. To confirm this applicability, we utilized the procedure to smooth a substrate with nano-striped corrugation pattern (Fig. 4.14c, d).

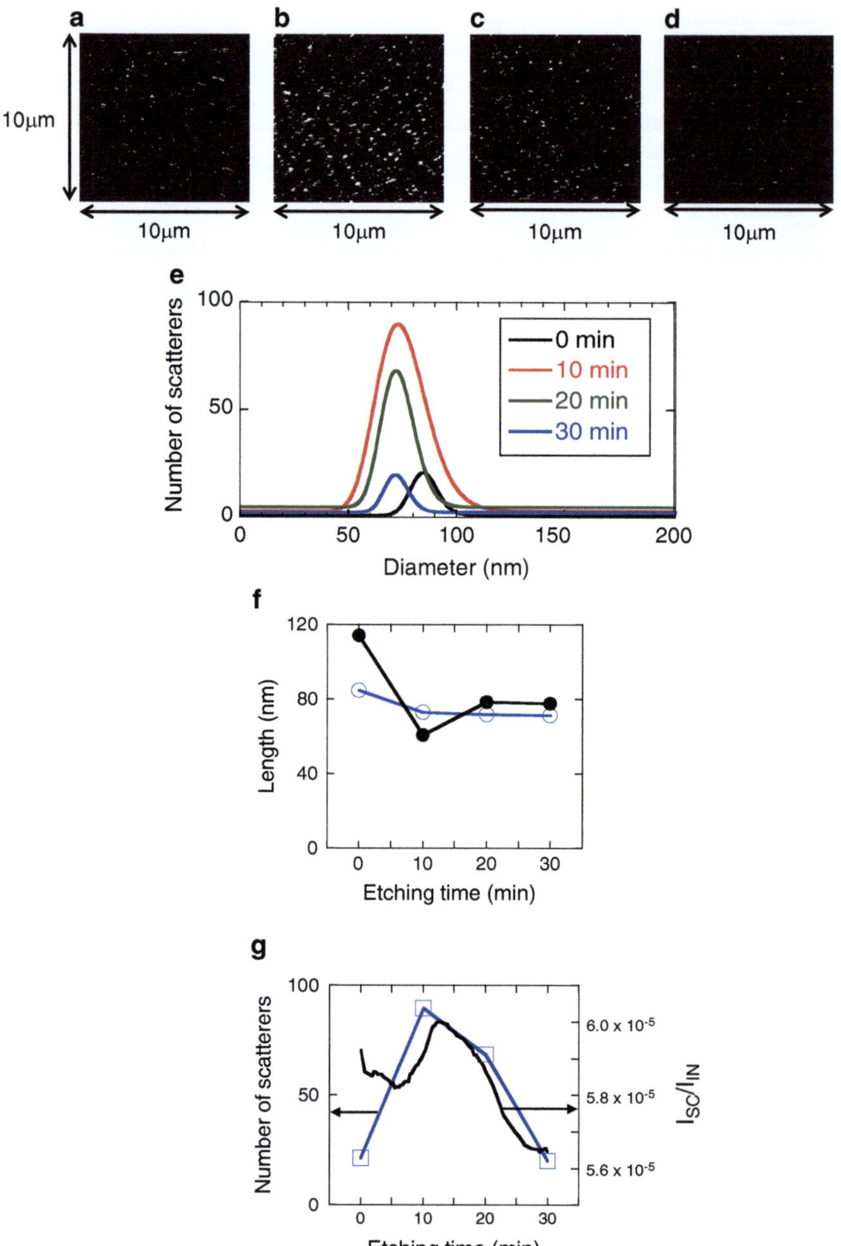

Fig. 4.13 (**a**)–(**d**) Digitized AFM image of etching time of 0 min, 10 min, 20 min, 30 min, respectively. (**e**) Scatterer diameter distribution for (**a**)–(**d**). (**f**) The comparison of L_c (same as the *solid circles* in Fig. 4.12a) and the obtained peak diameter of the scatterer. (**g**) Comparison of scatterer number and scattered light intensity I_{SC}/I_{IN} (previously plotted as the *solid line* in Fig. 4.10)

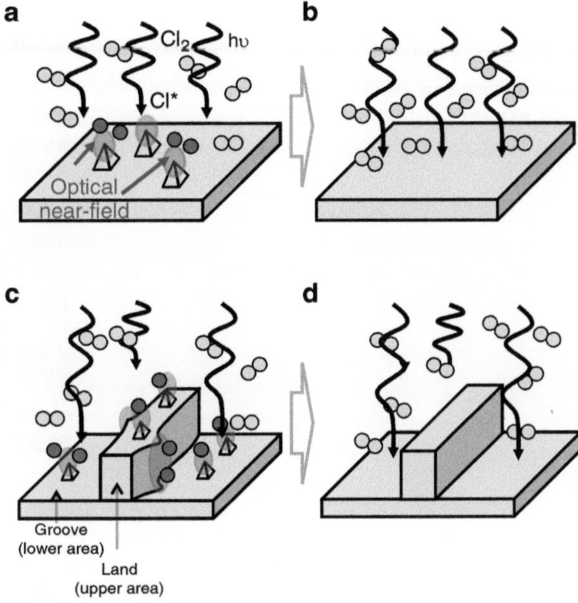

Fig. 4.14 Schematic diagram of near-field etching of (**a**) and (**b**) flat and (**c**) and (**d**) nano-striped substrates

The nano-stripe pattern was fabricated on of a soda-lime glass specimen, using a thermal nanoimprint technique with NiO molds [16]. The NiO mold patterns were transferred to the soda-lime glass at 600°C for 5 min under 10 MPa of pressure. Figure 4.15a shows a typical AFM image of the soda-lime glass with the nano-stripe corrugation pattern. The average height and period of the stripes were 13.5 nm and 175 nm, respectively.

Near-field etching was carried out by illuminating the substrate with a CW laser ($\lambda = 532$ nm) having a spatially uniform power density of 0.28 W/cm^2. The Cl$_2$ pressure in the chamber was maintained at 100 Pa at room temperature with a constant Cl$_2$ flow rate of 100 sccm (the same conditions used for smoothing a planner substrate of synthetic silica [11]). The surface morphology was evaluated via AFM after near-field etching. The scanning area of the AFM was 1.0 μm × 1.0 μm, and the scanned area was 256 (x-axis) ×256 (y-axis) pixels, with a spatial resolution of 4 nm for each axis.

Figure 4.15b shows a typical AFM image after 30 min of near-field etching. When this image is compared to Fig. 4.15a (before near-field etching), significant decreases in flank roughness can be seen. To evaluate the flank roughness reduction, we digitized the AFM images in order to analyze them numerically. Figure 4.16a, b shows the respective digitized AFM images of Fig. 4.15a, b; the white and black areas correspond to land and groove areas, respectively. The images were rotated to align them with the y-axis along the corrugations of the nano-stripe pattern. We also evaluated the land widths of white areas in the digitized AFM images. Figure 4.16c, d shows the respective land width w distributions before and after near-field etching, which were least-square fitted by the black solid Gaussian curves.

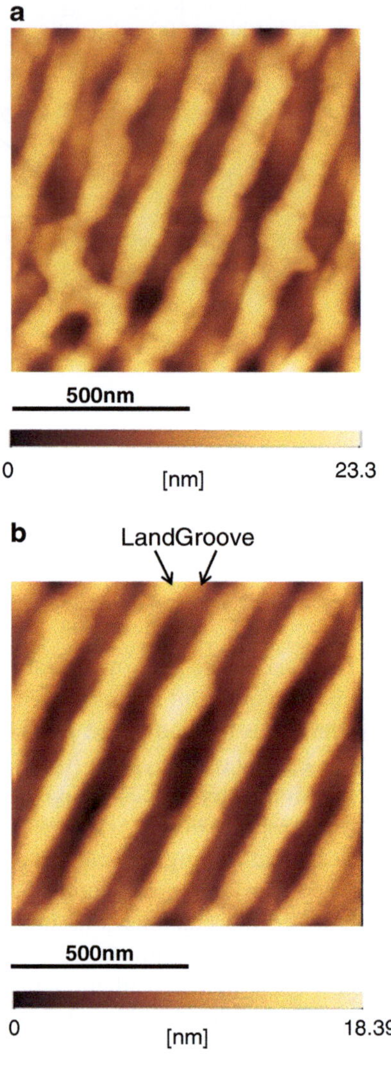

Fig. 4.15 Typical AFM images of the soda-lime glass (**a**) before and (**b**) after near-field etching (etching time 30 min)

By comparing these fitted curves, we found that after near-field etching, the central value of the land width w_c at the peak of the Gaussian curve decreased from 94.9 nm to 89.8 nm, and the standard deviation σ decreased from 20.7 nm to 17.6 nm. These simultaneous decreases in wc and σ indicate that near-field etching effectively reduced the flank roughness of the nano-stripe pattern. In addition to decreased flank roughness, a comparison of Fig. 4.16c, d confirms that land with width exceeding 125 nm disappeared, indicating that deburring is also realized by near-field etching. Furthermore, as Fig. 4.16d shows, the incidence of land with 90-nm width greatly exceeded the value of the fitted curve (the black solid curve), which also suggests that the deburring occurred (in other words, larger land was etched, and its width

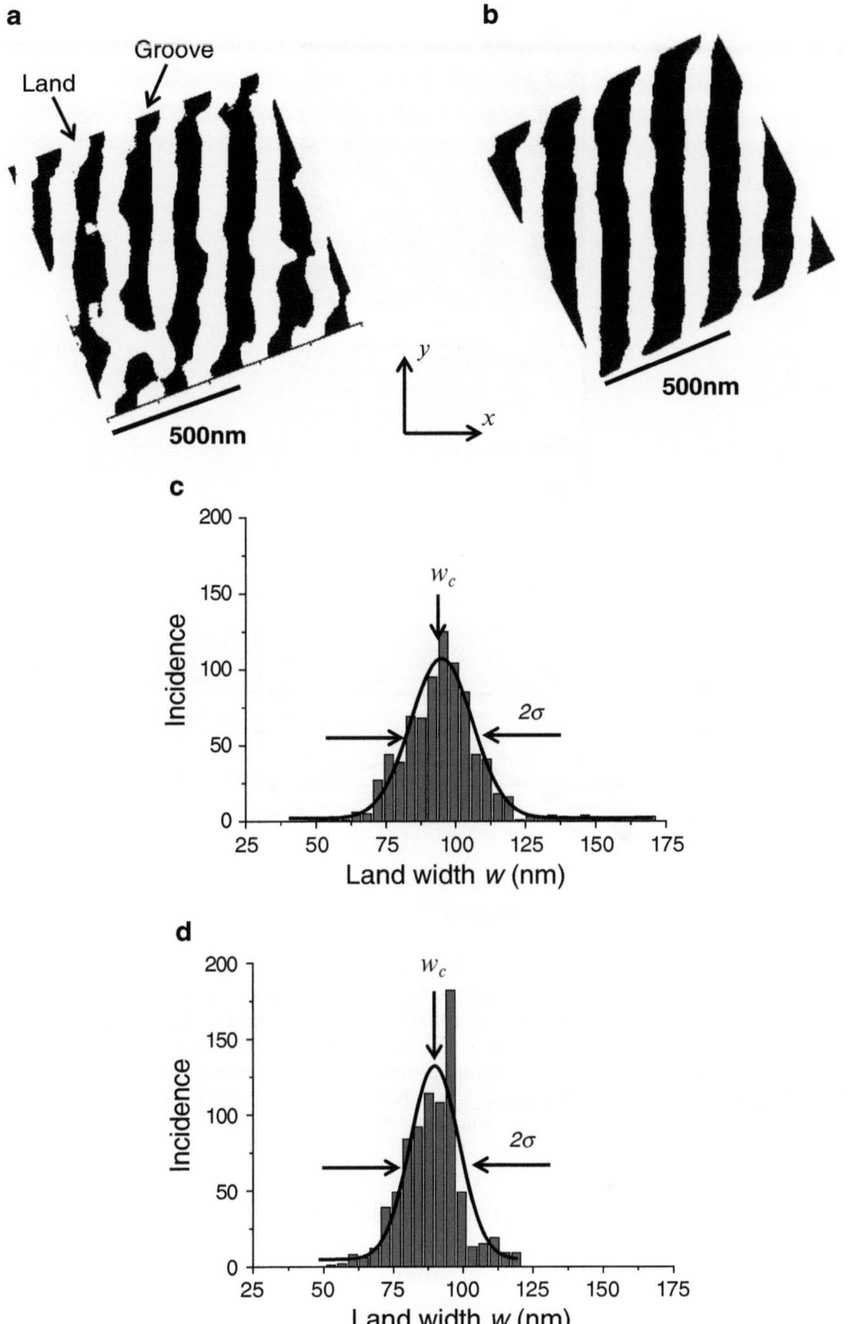

Fig. 4.16 Digitized AFM images (**a**) before and (**b**) after near-field etching. Land width distributions (**c**) before and (**d**) after near-field etching

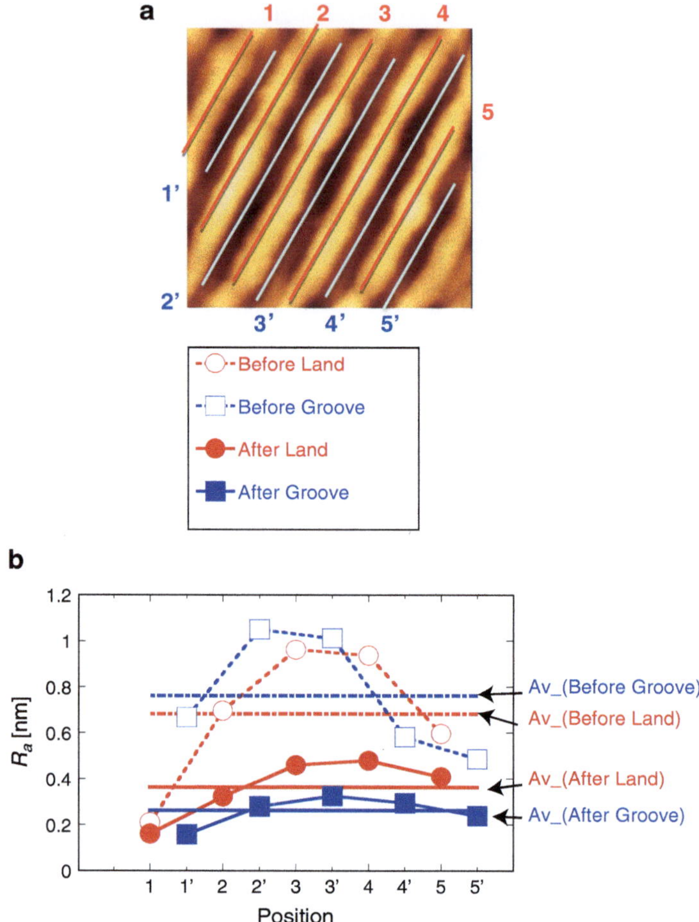

Fig. 4.17 (**a**) Schematic diagram of lines along land (*solid lines*) and grooves (*dashed lines*). (**b**) R_a values before near-field etching along land (*open circles* and *dashed line*) and grooves (*open squares* and *dashed line*). R_a values after near-field etching along land (*solid circles* and *solid line*) and grooves (*solid squares* and *solid line*). The number of positions corresponds to the number of the lines shown in (**a**). Average R_a: before etching of land: *dot-and-dashed line* (Av$_{\mathrm{BeforeLand}}$); after etching of land: *red solid line* (Av$_{\mathrm{AfterLand}}$); before etching of grooves: *dot-and-dashed line* (Av$_{\mathrm{BeforeGroove}}$); after etching of grooves: *blue solid line* (Av$_{\mathrm{AfterGroove}}$)

thereby decreased). Since the optical near field was selectively generated at the protrusions, selective etching of larger land was accomplished by near-field etching.

The roughness of the upper (land area) and lower (groove area) surfaces was also evaluated. We analyzed the surface roughness R_a along land and grooves (Fig. 4.17a). Figure 4.17b summarizes the R_a values before and after near-field etching. We found a significant reduction in R_a along each line; the average R_a for land decreased from 0.68 nm before etching (dot-and-dashed line) to 0.36 nm after etching (solid line), while the average R_a for grooves decreased from 0.76 nm before etching (dot-and-dashed line) to 0.26 nm after etching (solid line). Additionally, from the averaged values of R_a, the etching rates for land and grooves were estimated to be 0.64 nm/h and 1.0 nm/h, respectively, which were much larger than those of synthetic silica (0.14 nm/h) [11]. Since soda-lime glass has a longer absorption band edge wavelength (350 nm) [17] than synthetic silica (160 nm) [18], this higher etching rate originated of near-field generation caused by the higher absorption coefficient of soda-lime glass.

4.2.4 Repairing Nanoscale Scratched Grooves on Polycrystalline Ceramics Using Optical Near-Field Assisted Sputtering

As described in the previous section, the angstrom scale flatteness was realized using phonon-assisted etching. Since the phonon-assisted etching is the photo-chemical reaction, it can be applicable to other chemical reaction. In this section, the phonon-assisted chemical reaction can be applied to phonon-induced desorption which can realized repair of the scratches on polycrystalline Al_2O_3 ceramics.

Figure 4.18 shows a schematic of our method. As a result of preliminary polish of the Al_2O_3 ceramic, its surface contains nanoscale scratched grooves. Because the edges of the grooves have larger surfaces areas than the flat surface, the sputtered Al_2O_3 particles have a higher deposition rate at the edge after migration on the surface [19, 20]. As shown in Fig. 4.18a, it therefore is expected that Al_2O_3 will be deposited preferentially at the edges of scratched grooves, which will not help to repair the scratches. To avoid extra deposition on the edge of the grooves and to repair the scratches, we used the near-field desorption [21]. Because the optical near field (i.e., the dressed photon) can excite coherent phonons in the nanoscale structure, a virtual exciton-phonon-polariton (EPP) is generated on the substrate. A multistep transition via the EPP can accelerate the photochemical reaction although the photon energy is lower than the absorption band edge energy of the material [3]. When the ceramic is irradiated, a highly localized optical near field is generated at the edges of scratches, causing the photodesorption of depositing Al_2O_3 nanoparticles. As shown in Fig. 4.18b, if the light has a lower energy than the absorption band edge of the nanoparticles, effective deposition will decrease at the edge, and Al_2O_3 will accumulate on the bottom of the groove [21]. This process automatically stops after the scratches disappear, so that the optical near field can no longer be generated.

We performed optical near-field assisted sputtering to repair the scratches on the surface of the translucent Al_2O_3 ceramic SAPPHAL ® [22]. Figure 4.19 shows a schematic of the experimental setup. Planar surfaces of SAPPHAL ® substrate were prepared by polishing using diamond abrasive grains with a diameter of 0.5μm. The Al_2O_3 was deposited using radio-frequency (RF) magnetron sputtering (RF power: 300 W; frequency: 13 MHz). The total gas pressure was 7×10^{-1} Pa, with a gas flow of 16 sccm Ar and 1.2 sccm O_2 [23]. We also used SAPPHAL

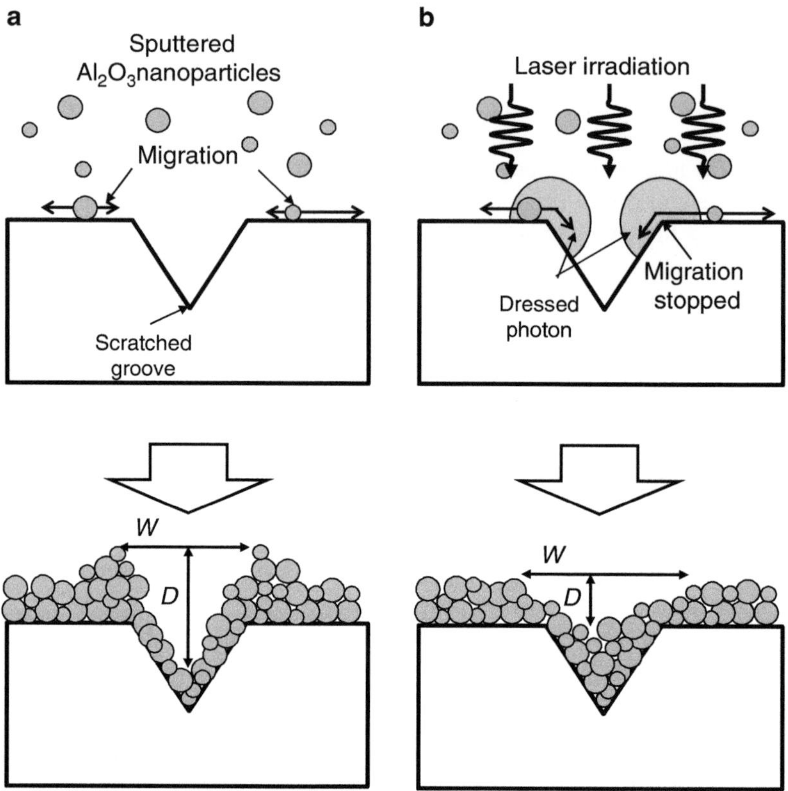

Fig. 4.18 Schematics of deposition on the scratched substrate surface (**a**) without and (**b**) with irradiation during the sputtering. D is the depth and W is the width of the scratched groove

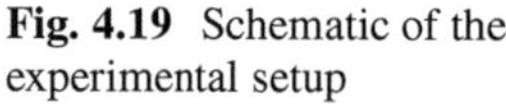

Fig. 4.19 Schematic of the experimental setup

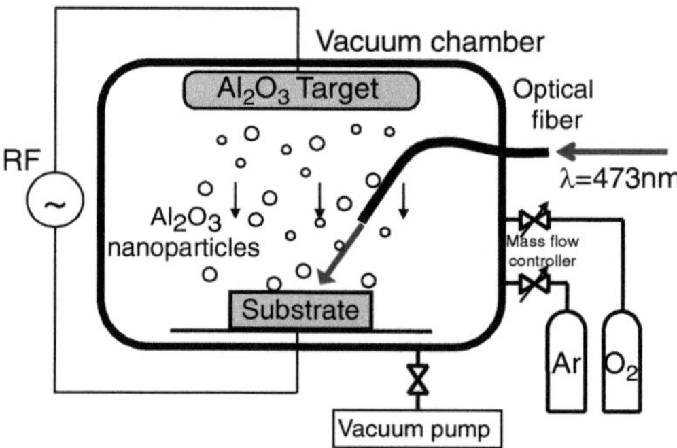

as the target material for sputtering. The CW second harmonic of a Nd:YAG laser with wavelength of $\lambda = 473$ nm was used as the light source for optical near-field generation. During the sputtering process, the surface was irradiated with light with an optical power density of 2.7 W cm^{-2}. As stated previously, the photon energy of

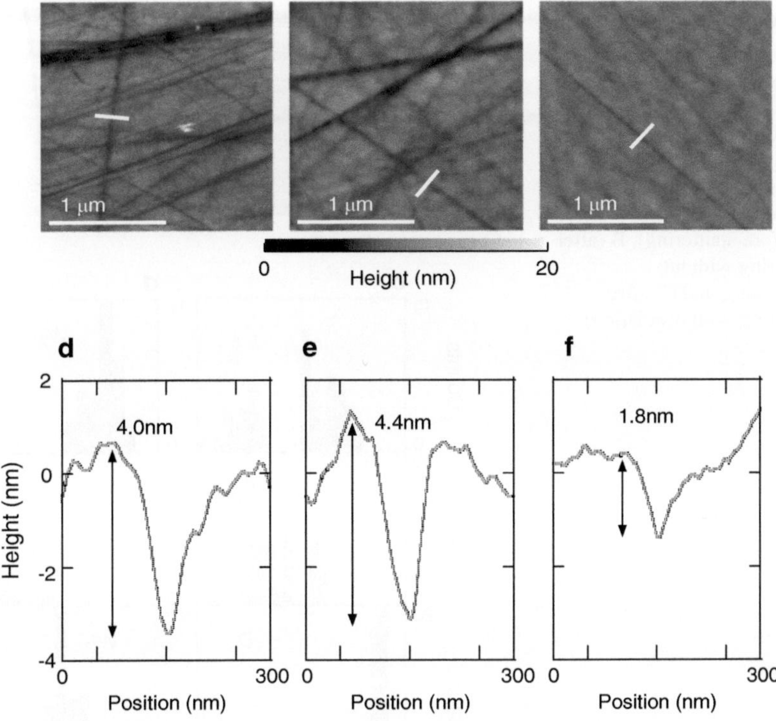

Fig. 4.20 AFM images of substrates (a) before sputtering (substrate A), (b) after sputtering without irradiation (substrate B), and (c) after sputtering with irradiation (substrate C). (d), (e), and (f) Cross-sectional profiles along the *white dashed lines* in (a), (b), and (c), respectively

this laser was lower than the absorption band edge energy of Al_2O_3 ($\lambda = 250$ nm) [24]. The light was introduced to the ceramic surface through a multimode optical fiber. After 30 min of sputtering, the thickness of the deposited Al_2O_3 layer was about 100 nm. We observed the surface of the substrates using an AFM.

Figure 4.20a–c shows typical AFM images of SAPPHAL ® surfaces before sputtering, after sputtering without irradiation, and after sputtering with irradiation, respectively. Figure 4.20d–f shows the cross-sectional profiles of typical scratches along the white dashed lines in Fig. 4.20a–c, respectively. The depth of the scratched grooves in Fig. 4.20e is 4.4 nm, which is deeper than that in Fig. 4.20d (4.0 nm). This was caused by the extra deposition at the edges of the scratch due to the low surface potential, as described in Fig. 4.18a. In contrast, Fig. 4.20f shows that the depth decreased to 1.8 nm without extra deposition at the edges of scratches after sputtering with irradiation. For more quantitative evaluation, we calculated the surface roughness Ra. The R_a values over the AFM images of Fig. 4.20a–c were $R_{aA} = 1.3$ nm, $R_{aB} = 1.1$ nm, and $R_{aC} = 0.49$ nm, respectively. These results indicated that the repair of scratched grooves by optical near-field desorption at the edge of the scratches resulted in a drastic decrease in the surface roughness.

Fig. 4.21 (**a**) An AFM image is leveled by a least square method. (**b**) Schematic of the Hough transform. *Straight lines* of the scratches are detected using the Hough transform of (**a**). (**c**) Histograms of depths, D_n. (**d**) widths (W_n) of scratches on alumina ceramics substrates; A (before sputtering), B (after sputtering without irradiation), and C (after sputtering with irradiation)

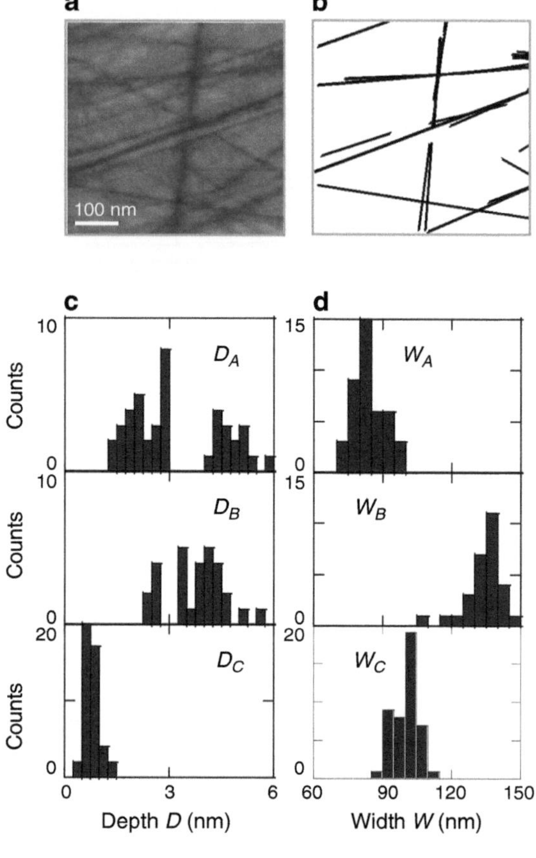

To selectively evaluate the profiles of the scratched grooves, we used the Hough transform [25]. After the AFM image of Fig. 4.21a was leveled by a least square method and binarized, the linear features of scratches were automatically extracted using a Hough transform; see Fig. 4.21b. Through this method we obtained the depth D and width W of the detected scratches. Figure 4.21c, d shows the statistical analyses of D and W obtained from the images in Fig. 4.20a–c, respectively. As shown in Fig. 4.21c, the average values of D were $\overline{D_A} = 3.2\,\text{nm}$, $\overline{D_B} = 3.8\,\text{nm}$, and $\overline{D_C} = 0.79\,\text{nm}$, respectively, which confirmed that the depth of the scratches were drastically decreased using the near-field assisted sputtering. Figure 4.21d shows that the width of the scratches after the sputtering without irradiation (W_B) was increased beyond the original value W_A, which also supports our deposition model; see Fig. 4.18a. In addition, the width W_C also increased in comparison to W_A, supporting the model of Fig. 4.18b. Further decreases in the width could be achieved by optimizing laser and sputtering conditions.

4.3   Site Selective Patterning

Nanometer-sized photonic integrated circuit has been proposed for the use in future optical transmission and signal processing systems with high data transmission rates and capacity [26]. Novel nanophotonic AND-gate [27], and NOT-gate devices [28], and nano-fountains [29] have been reported, and their operation was demonstrated by controlling the optical near-field energy transfer in quantum cubes and quantum wells. To fabricate these devices, with sub-10-nm quantum dots and wires, sizes and positions must be controlled with nanometer-scale accuracy. To realize this degree of accuracy, electron beam lithography [30] and scanning probe microscopy [31] have been applied, but these techniques have a low throughput and are not suitable for production.

4.3.1   Production of Size-Controlled Si Nanocrystals Using Self-organized Optical Near-Field Chemical Etching

To realize efficient production of nanophotonic devices, we propose a self-organizing method of photochemical etching, in which a high degree of size-control is achievable using size-selective excitation due to quantum-size effects [32, 33]. As the photochemical etching proceeds, the size of Si nanocrystals decreases, and the absorption band gap energy of the Si increases due to the quantum size effect. Finally, the photochemical etching stops when the size of Si nanocrystallite decreases, and the absorption edge of the particles shifts to higher energies, owing to the confinement of the excitons. The photochemical etching stopes when the Si nanocrysltas become so small that they no longer absorb the illumination.

We fabricated a template of the pyramidal Si structure using photolithography and anisotropic etching of (100)-oriented Si to generate the optical near field. The pyramidal Si structure were fabricated by anisotropic etching (40g: KOH + 60 g: H_2O + 40 g: isopropyl alcohol, 80°C) (Fig. 4.22a) [34, 35]. Photochemical etching was then carried out using a solution of HF (50%): H_2O_2 (30%) = 1:1 at room temperature, and under illumination of a laser with a photon energy of 1.8 eV (26 mW/cm^2, beam diameter of 500 μm). The photochemical etching proceeds as follows [36]:

1. Laser irradiation forms an electron–hole pair in the Si (Fig. 4.22b).
2. Si atom is oxidized by H_2O and holes.
3. The SiO_2 region is dissolved by HF.
4. H_2O_2 removes electrons from the substrate, and H_2O_2 and H^+ react to form water (Fig. 4.22c).
5. The remaining Si results in rod-like bridging nanocrystals (Fig. 4.22d) [37].

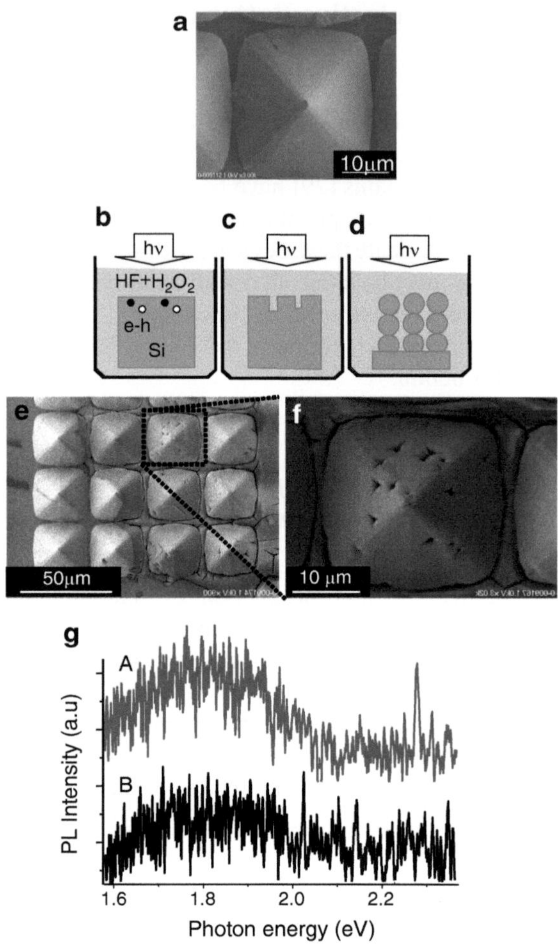

Fig. 4.22 (**a**) SEM images of pyramidal Si. (**b**)–(**d**) A schematic of photochemical etching of Si substrate and the formation of bridging Si nanocrystals. (**d**) SEM images of pyramidal Si after dipping in the etching solution for 3 min with illumination. (**e**) Magnified image of the *dashed square* in (**d**). (**f**) Room temperature PL spectra of the photochemically etched Si illuminated for 3 min (curve A) and 10 s (curve B)

A half wave plate was used to examine the polarization dependence of the etching, and the etched surface morphology was measured by scanning electronic microscopy (SEM).

Although no change in surface morphology was observed after 30 min of etching without illumination, the surface became darker under illumination for 3 min as shown in Fig. 4.22e, f, in which the right-hand side area of the pyramid was illuminated. To identify the origin of this change, we measured room temperature photoluminescence (PL) using a continuous wave He–Cd laser (with a photon energy of 3.81 eV). As shown by the curve A in Fig. 4.22g, the PL has a spectral peak at 1.8 eV, which corresponds to 2.8-nm-diameter Si nanocrystals [32]. Although the spatial resolution of the SEM image may not be sufficiently high for determining the grain size of Si nanocrystallites accurately, note that the dark area of the surface was etched photochemically to produce Si nanocrystals that were 2.8 nm in diameter.

Furthermore, because the energy of this spectral peak energy is equal to the photon energy used for the etching, the size of the Si nanocrystals was controlled by the photon energy of the light source.

Since the photon energy used for the etching was larger than the absorption edge energy of Si (1.1 eV), only the area inside the beam spot was etched. Figure 4.23a, b shows the SEM images after 10 s of etching with the incident light polarized 45 degree (P_{45}) and in parallel (P_0) to the x-axis, respectively. P_{45} resulted in two dark ridges along the polarization, which are encircled by broken ellipses in Fig. 4.23a. However, the polarization P_0 resulted in four dark ridges, which are highlighted by the broken ellipses in Fig. 4.23b.

Figure 4.23d shows the spatial distribution of the near-field PL intensity at 1.8-eV photon energy (curve B in Fig. 4.22g shows typical PL spectrum of Fig. 4.23b), imaged by scanning an apertured probe with an aperture diameter of 200 nm. From

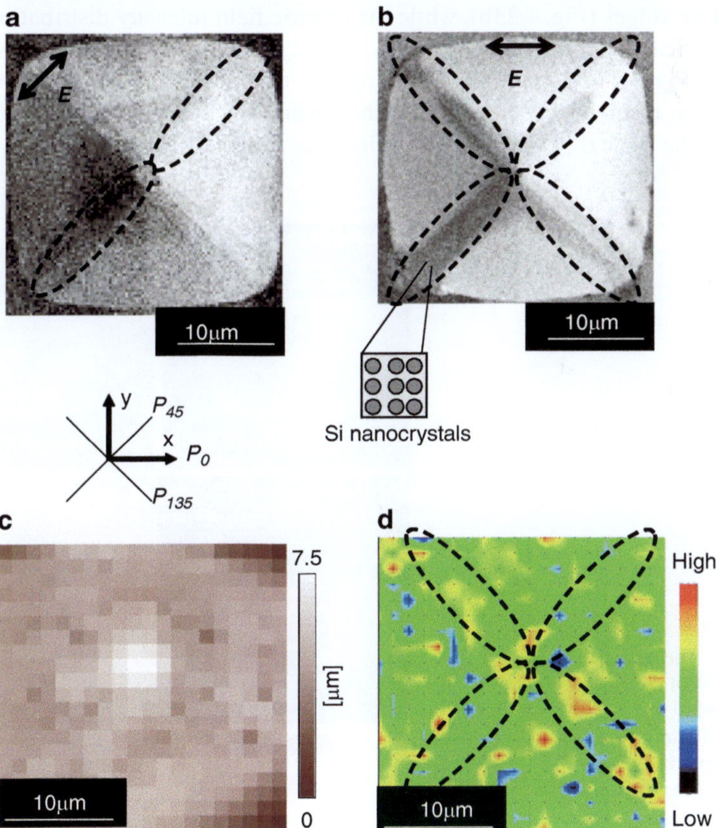

Fig. 4.23 SEM images of the surface morphology after 10 s photochemical etching with polarization (**a**) P_{45} and (**b**) P_0. (**c**) Topographical image of the pyramidal Si. (**d**) Spatial distribution of near-field PL intensity at 1.8 eV

this figure, we find that PL originated from the ridges of the pyramidal Si along $45°$ and $135°$ azimuth angles. These results confirm that the dark area in Fig. 4.23a was selectively etched, resulting in Si nanocrystals with a 2.8 nm diameter.

Although the polarization P_{45} produced Si nanocrystals along the two ridges of the $45°$ azimuth angle (Fig. 4.23a), the polarization P_0 resulted in Si nanocrystals on four ridges (Fig. 4.23b). To explain such polarization-dependent etching, we calculated the optical field intensity of the pyramidal Si, in which the four sidewalls resulted from the (111) silicon crystal planes, using a FDTD method. Figure 4.24a, b shows the spatial distributions of the electrical field intensity, $I(r)$, at a height of 10 nm over the Si tip (see Fig. 4.24e) with polarizations P_{45} and P_0, respectively. Figure 4.24c, d shows $I(r)$ at the 250 nm from the Si tip with polarization P_{45} and P_0, respectively. Figure 4.24c reveals strong optical field intensity at the corners B and D. It is therefore quite reasonable to selectively etch the ridges of the $45°$ azimuth angle via the polarization P_{45} as shown in Fig. 4.23a.

At first inspection, the polarization P_0 resulted in the production of Si nanocrystals on four ridges (Fig. 4.23b), while the electric field intensity distribution reveals strong optical field intensity not only at the corners but also along sides AB and DC (Fig. 4.24d). Therefore, they appear to contradict each other. However, this can be resolved by considering the virtual exciton-phonon-polariton (EPP) model of the optical near field [1–4]. Since the dressed photon can excite coherent

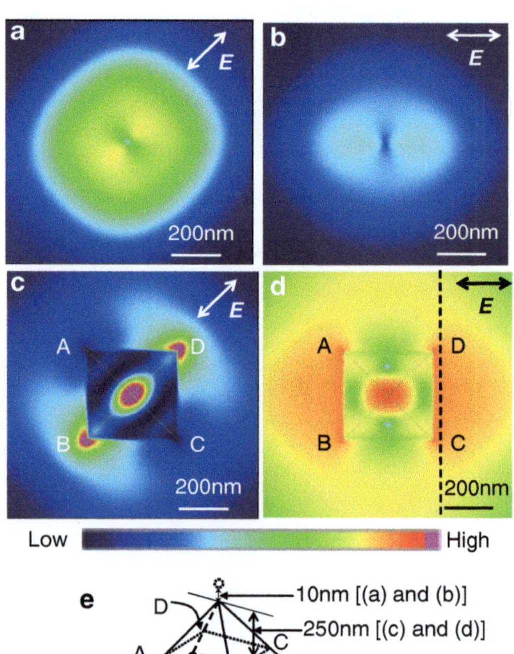

Fig. 4.24 The spatial distribution of the electrical field intensity, $I(r)$, at a height of 10 nm above the Si tip with polarization (**a**) P_{45} and (**b**) P_0, respectively. The spatial distribution of the electrical field intensity $I(r)$ 250 nm below the Si tip with polarization (**c**) P_{45} and (**d**) P_0, respectively. (**e**) Schematic of the pyramidal Si

phonons in the nanoscale structure, a virtual EPP is generated on the Si surface. A multistep transition via the EPP can also excite charge carriers, accelerating the photochemical reaction. The spatial distribution of the virtual EPP, and thus the rate of the photochemical reaction, can be represented by the product of the optical field intensity, $I(r)$, and its gradient $dI(r)/dr$ [38]. Therefore, we calculated $INF = I(r) \times dI(r)/dr$, which depends on the photochemical reaction originating from the virtual EPP. Figure 4.25a, b shows the FDTD simulated intensity distribution of with P_{45} and P_0, respectively, where

$$\frac{dI(r)}{dr} = \sqrt{\left\{\frac{\partial I(x,y,z)}{\partial x}\right\}^2 + \left\{\frac{\partial I(x,y,z)}{\partial y}\right\}^2 + \left\{\frac{\partial I(x,y,z)}{\partial z}\right\}^2}. \qquad (4.9)$$

In the simulation, a cell size of 5 nm was used, together with a photon energy of 1.8 eV. The refractive index of the Si was 3.8 [39]. Comparison with the cross-sectional profile $I(r)$ in Fig. 4.25c, which was taken along the dashed line in

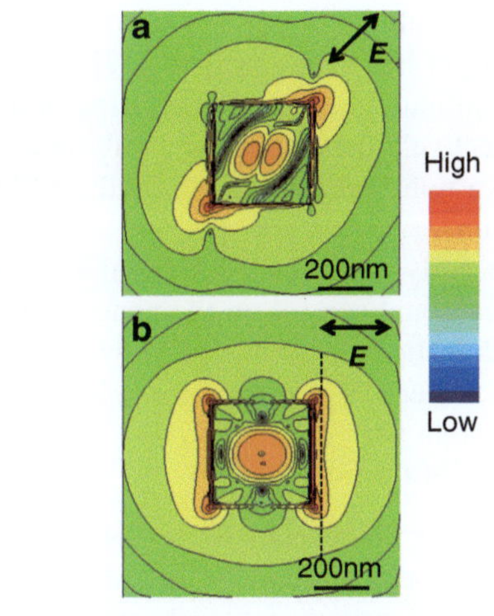

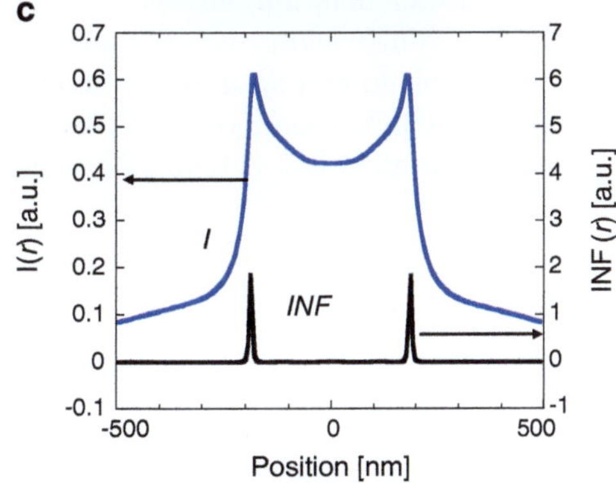

Fig. 4.25 The spatial distribution of the optical near-field component $I(r) \times dI(r)/dr$ at the 250-nm below the Si tip with polarization (**a**) P_{45} and (**b**) P_0, respectively. (**c**) Cross sectional profiles along the *dashed lines* in Fig. 4.24d (curve I) and Fig. 4.25b (curve INF)

Fig. 4.24d, reveals a strongly localized distribution of the virtual EPP at the corners along the dashed line in Fig. 4.25b shown by the curve *INF* in Fig. 4.25c. The calculated distribution of the virtual EPP was in good agreement with the spatial location of the etching with polarization P_0 (resulting in four dark ridges, as shown in Fig. 4.23b).

4.3.2 Site-Selective Deposition of Gold Nanoparticles Using Non-adiabatic Reaction Induced by Optical Near Fields

Metal nanoparticles have been attracting attention as materials for nanoscale photonic and electronic devices [40–43]. Gold nanoparticles in particular have been studied extensively because of their chemical stability [44]. There are several methods of producing gold nanoparticles, such as sputtering [45] and photochemical synthesis [46]. Among these methods, photochemical fabrication using $AuCl_4^-$ ions has the advantage of deposition-site selectivity by selecting the light irradiation area. However, although positional control of gold nanoparticles using interference of laser light has been reported [47], positional control on the nanometric scale has been difficult because of the diffraction limit of light. In this study, we succeeded in selectively depositing gold only in an area below the optical diffraction limit by using a nonadiabatic reaction induced by optical near fields as the photochemical reaction.

Recently, methods involving the use of optical near fields have been considered for positional control on the nanometric scale using light [26]. In addition, with a nonadiabatic process mediated by coherent phonons induced close to the surface of a nanomaterial by optical near fields, it is possible to induce novel reactions that are impossible with conventional approaches involving propagating light [3]. Typical features of this method are:

1. Because the reaction is mediated by coherent phonons, it is possible to induce a photochemical reaction in the material even when using light having a photon energy lower than the absorption edge of the material. As a result, a short-wavelength light source is not required.
2. It is possible to process a nanostructure on the material surface to function as a nanometric light source simply by using propagating light, eliminating the need for a probe and mask [11, 48]. As a result, processing of large surface areas is possible.

Here we explain deposition of gold particles by a nonadiabatic process, which is the principle of the present study. For comparison, we also explain an adiabatic process that has often been used conventionally. We consider a case where a nanostructure is placed in contact on top of a film containing $AuCl_4^-$ serving as a source of gold particles. When this nanostructure is irradiated with light, optical near fields are generated around the nanostructure. If the photon energy of the irradiated

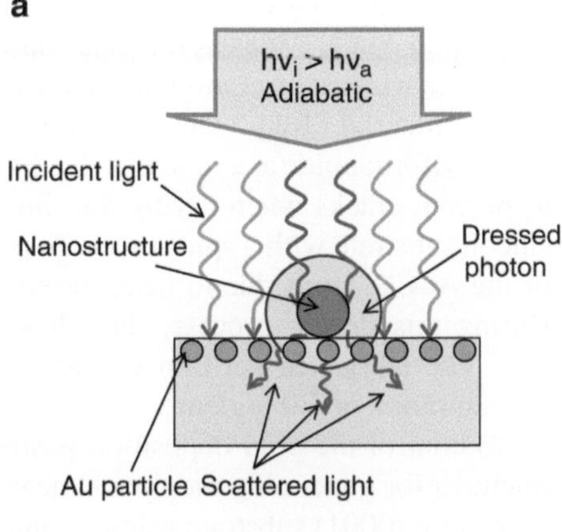

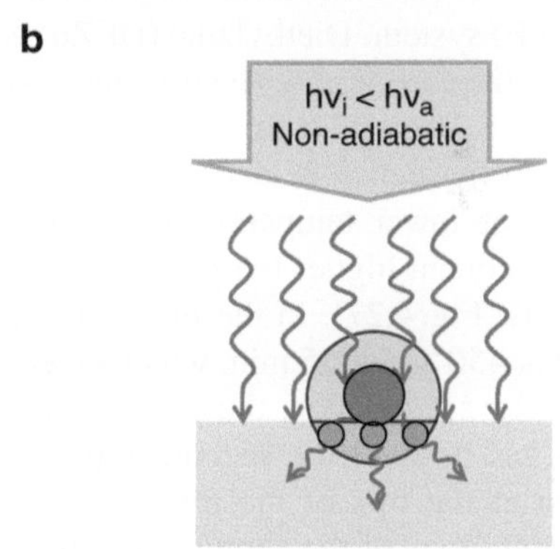

Fig. 4.26 Schematic diagram of photodeposition of gold nanoparticles using optical near fields. (**a**) Adiabatic process, (**b**) nonadiabatic process

light ($h\nu_i$) is greater than the photoreduction energy of the $AuCl_4^-$ ions ($h\nu_a$), gold nanoparticles are deposited via a normal photochemical reaction in the entire irradiated region by the incident light and scattered light (Fig. 4.26a). This reaction process is called an adiabatic process. In contrast, if $h\nu_i < h\nu_a$, it is normally not possible to bring about a photoreaction, and thus no gold particles are deposited in the area irradiated with incident light and scattered light. However, optical near fields are generated close to the surface of the nanostructure, and gold nanoparticles are deposited by these optical near fields even if $h\nu_i < h\nu_a$ (Fig. 4.26b). This is because the optical near fields can induce a reaction with the $AuCl_4^-$ ions by binding with coherent phonons. This reaction process is called a nonadiabatic process and occurs only around the nanostructures which generate the optical near fields. By using the nonadiabatic process, it is possible to eliminate the influence of propagating light, that is, incident light and scattered light, allowing only the optical near-field effect to be utilized [11, 48]. In this paper, we report on the formation of gold nanoparticles using optical near fields, which enables positional control on the nanometric scale.

In order to achieve deposition of gold nanoparticles, we used a silica gel film containing $AuCl_4^-$ ions as the gold source. First, we fabricated silica sol. The silica sol was fabricated by combining $HAuCl_4$ ethanol solution (0.5 g/cm^3), spin-on glass (SOG) material (Tokyo Ohka Kogyo Co., Ltd., Tokyo, Japan; OCD15000-T), and dimethylformamide at a ratio of 1:2:1 by volume. The dimethylformamide was used to prevent cracks when drying the film. A film of the silica sol was coated on a silicon substrate with a spin coater. To prevent progression of the reduction reaction of the $AuCl_4^-$ ions due to heat, the film was vacuum dried at room temperature. During this drying process, the silica sol film was transformed to a silica gel film. The fabricated gel film was about 200 nm thick and contained $AuCl_4^-$ at a concentration of 6.0 mg/cm^3.

To control the gold deposition position, we used ZnO nanorods as the nanostructures for generating the optical near fields [49]. ZnO nanorods were grown on a sapphire (0001) substrate using a catalyst-free metal-organic vapor phase epitaxy (MOVPE) system. Diethylzinc (DEZn) and oxygen were used as the reactants, with argon as the carrier gas. The pressure inside the reactant chamber was maintained at 5 Torr. The substrate temperature was controlled using a thermocouple thermometer and radio-frequency-heated carbon susceptor. Our two-temperature growth method consists of lower temperature growth of vertically aligned thick ZnO nanorods and subsequent higher temperature growth to fabricate vertically aligned ultrafine nanorods (Fig. 4.27). In the first step, vertically aligned thick ZnO nanorods were grown at 450°C for 35 min, which determined the growth direction in the next step. Before the second step, the substrate temperature was increased to 750°C without DEZn gas flow. In the second step, ultrafine nanorods were grown at 750°C for 10 min at the tips of the preformed thick nanorods. SEM was used to observe the morphology of the nanorod. For further investigation, we obtained transmission electron microscopy (TEM) images and selection area diffraction (SAD) patterns. The optical properties were evaluated using PL measurements.

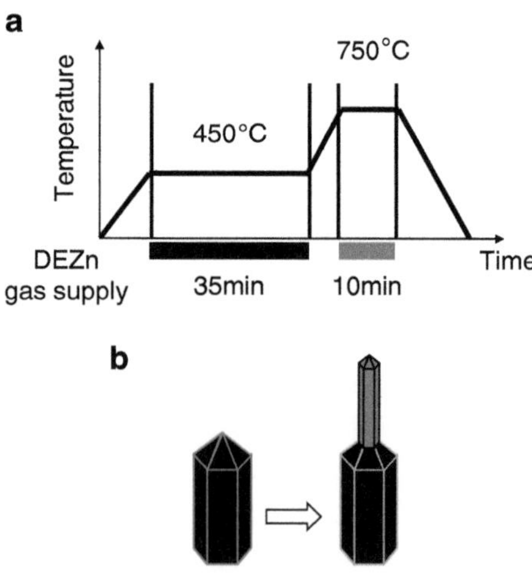

Fig. 4.27 (a) Schematic representation of the two-temperature growth method. (b) Schematic illustration of nanorod growth

Fig. 4.28 SEM images of
ZnO nanorods grown at
constant temperatures of (**a**)
450°C and (**b**) 750°C. (**c**)
and (**d**) SEM images of
vertically aligned ultrafine
nanorods grown using the
two-temperature growth
method

Figure 4.28a, b shows the SEM images of ZnO nanorods fabricated at constant growth temperatures of 450°C and 750°C for 60 min, respectively. The morphology of the ZnO nanorods depend on the growth temperature. We grew vertically aligned ZnO nanorods with a mean diameter of 100 nm at 450°C. In contrast, the ZnO nanorods grown at 750°C were oriented in random directions, but had tip diameters as small as 20 nm. The temperature dependence of the radius is in good agreement with a previous report [50]. Figure 4.28c, d presents the SEM images of ZnO nanorods grown in the second step, showing ultrafine ZnO nanorods grown on the tips of thick preformed nanorods. The direction of the ultrafine ZnO nanorods was vertical to the substrate. The mean diameter of the ultrafine ZnO nanorods measured

from SEM images was 17.7 ± 3.3 nm (mean $\pm$ SD). The ultrafine ZnO nanorods obtained in the second step should show radial quantum confinement.

We used TEM to investigate the structural characteristics further. A low-magnification TEM image (Fig. 4.29a) revealed the vertical growth of ultrafine nanorods on the sapphire substrate. A high-resolution TEM image of an ultrafine nanorod tip (Fig. 4.29b) and corresponding SAD pattern (Fig. 4.29c) revealed that the ZnO nanorods consisted of single crystals with a lattice spacing of $a = 0.33$ nm and $c = 0.51$ nm. These values agree with those of crystalline wurtzite ZnO ($a = 0.3249$ nm and $c = 0.5207$ nm) within the measurement error. The TEM measurement also revealed that the nanorods grew along the c-axis ([0001] direction). Figure 4.29d shows a high-resolution TEM image of the border between the thick and ultrafine nanorods grown in the first and second steps, demonstrating that the crystal alignment is maintained in both steps. We calculated the growth rate in each step from the ratio of the length of the fabricated nanorods to the growth time, which were 37 nm/min and 130 nm/min in the first and second steps,

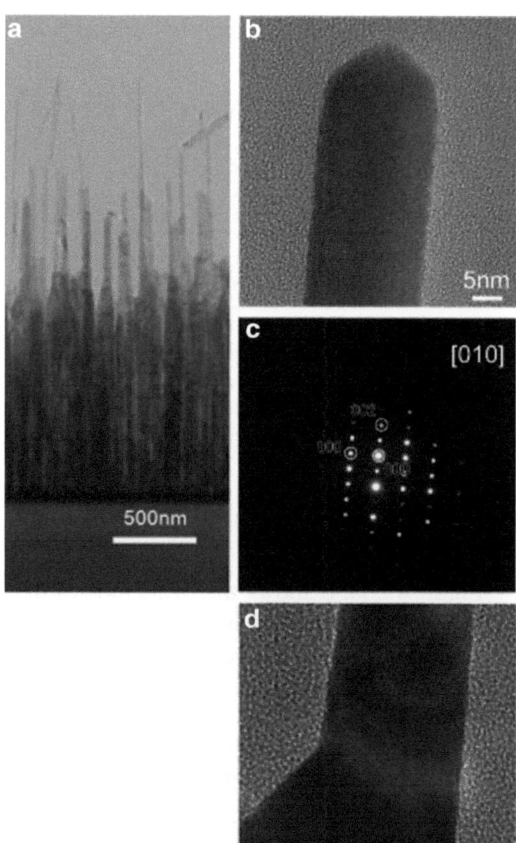

Fig. 4.29 (a) Low-resolution TEM image of the vertically aligned ultrafine nanorods. (b) High-resolution TEM image of the tip of an ultrafine nanorod and (c) the corresponding SAD pattern. (d) High-resolution TEM image of the border between the thick and ultrafine nanorods

respectively. The increased growth rates and decreased diameters at the higher growth temperature can be explained using anisotropic surface energy. Since the surface energy of the (0001) plane is greater than that of the other crystal faces, such as (01-10) [51–53], growth along the [0001] direction was favored energetically [50]. As the growth temperature increases, the migration of the atoms adsorbed to the surface is enhanced, from low-surface energy (01-10) planes to high-surface energy (0001) planes. Therefore, a high growth temperature results in high growth rates in the [0001] direction and smaller realized diameters.

We acquired PL spectra to evaluate the optical properties of the ZnO nanorods. The 325-nm line from a continuous-wave He–Cd laser was used as the excitation source. The shutter speed was 0.01 s for the range 5–40 K, 1 s for the range 60–150 K, and 10 s at 300 K. The ZnO nanorods exhibited PL emission peaks at between 5 K and 300 K (Fig. 4.30). A low temperature (5 K) PL spectrum had peaks and shoulders at 3.319 eV, 3.369 eV (I_2), and 3.381 eV (I_{ex}). The peaks at 3.369 eV and 3.319 eV corresponded to the emission from the neutral-donor bound exciton and respective two-electron transition [54]. The shoulder at 3.381 eV corresponded to the emission from the free exciton. The emission from the free exciton supports the high crystal quality of these nanorods. The intensity of peak I_2 decreased as the temperature increased, and peak Iex became dominant. This behavior presumably resulted from the decomposition of bound excitons into free excitons due to the increased thermal energy, and supports the above argument. The full-width at half-maximum of peak I_2 was 8.4 meV and 86 meV at 5 and 300 K, respectively. In the PL measurement, since the collected PL spectra measured an

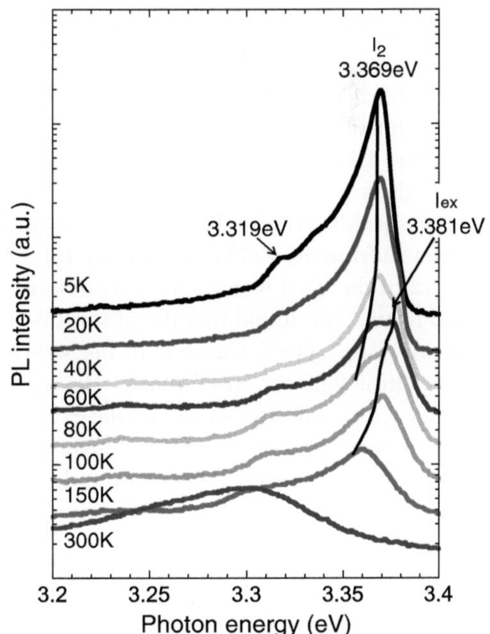

Fig. 4.30 Temperature dependence of the PL spectra of ZnO nanorods grown with the two-temperature growth method

Fig. 4.31 (a) Experimental setup. (b) Absorption spectra of ZnO nanorods (curve A) and HAuCl$_4$ solution (curve B), and photoluminescence (PL) spectrum of ZnO nanorods (curve C)

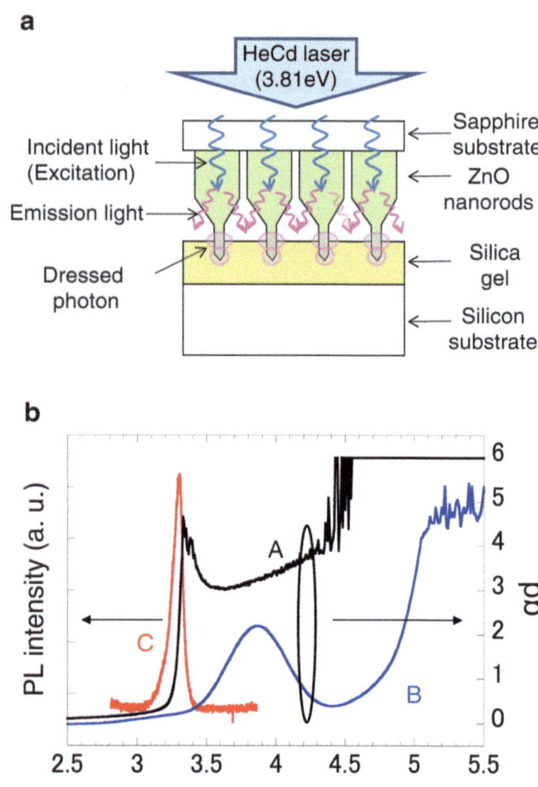

ensemble of thousands of ZnO nanorods, we could not distinguish between the emission from thick nanorods and that from ultrafine nanorods.

Figure 4.31a shows the experimental setup. The substrate on which the ZnO nanorods were grown was pressed into the silica gel film containing the AuCl$_4^-$ and was irradiated with He–Cd laser light (3.81 eV, 15 mW, 1 mm irradiation spot diameter) from the ZnO nanorod side for 1 min. Figure 4.31b shows the absorption spectra of the ZnO nanorods (curve A) and the HAuCl$_4$ solution (curve B). From these results, we found that the absorption edge energy of the AuCl$_4^-$ ions in the solution was 3.44 eV. Curve C in Fig. 4.31b is the ZnO nanorod PL spectrum due to excitation with the He–Cd laser. The peak wavelength of the PL spectrum was at 3.30 eV. These findings show that the excitation light was absorbed by the first-stage ZnO nanorods and did not reach the silica sol film. In addition, the photon energy of light emitted from the ZnO nanorods was smaller than the absorption energy of the AuCl$_4^-$ ions, making it impossible to achieve gold deposition via the adiabatic process. Therefore, in this experiment, the gold deposition occurred only via the nonadiabatic process by light emission from the ZnO nanorods.

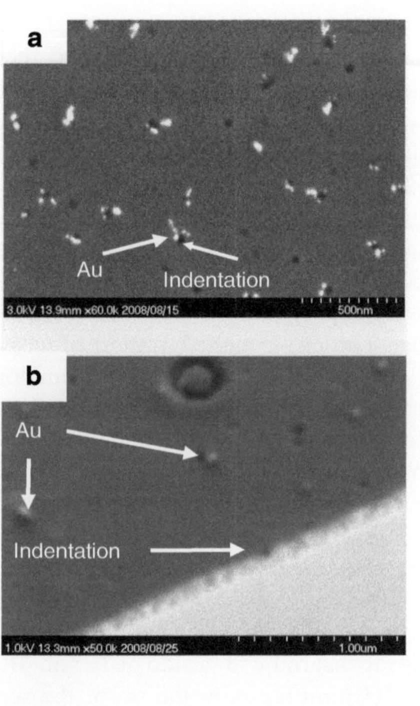

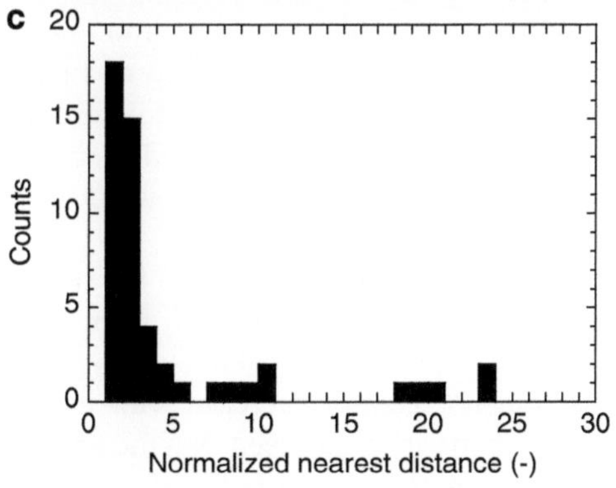

Figures 4.32a, b is a top-view and perspective-view SEM images of the surface of the silica gel film after the ZnO nanorods were pressed into it and irradiated with light. The white points indicated by the solid-line arrows show the deposited gold nanoparticles, and the dark points indicated by the broken-line arrows show indentations of traces left by pressing the ZnO nanorods. The surface density of the ZnO nanorod traces was smaller than the surface density of the ZnO nanorods themselves. This is because of some variation in the lengths of the ZnO nanorods employed, meaning that only ZnO nanorods with a large height were pressed into the silica gel. Also, the sites where deposited gold particles were produced

were sparse despite the fact that the spot irradiated by the He–Cd laser light was 1 mm in diameter. We evaluated the gaps between the gold nanoparticles and the nanorod traces. The shapes of the gold nanoparticles and the ZnO nanorods were approximated by circles of equal area. Figure 4.32c shows the distribution of distances between the centers of gravity of the gold nanoparticles and the nearest ZnO nanorod traces, normalized by the radii of the ZnO nanorod traces. Regarding the distances between the centers of gravity, 36% of the total were 2–3 times greater than the radii of the nanorod traces, and 30% were 3–4 times greater; thus, considering that the actual shape is non-circular, it means that the gold nanoparticles were deposited adjacent to the nanorod traces. The average diameter of the gold nanoparticles obtained from SEM measurements was 17.7 ± 6.1 nm.

We performed numerical calculations to show that the deposition of gold only in the vicinity of the nanorods was due to the nonadiabatic process. A nonadiabatic reaction involving excitation of coherent phonons is described by the virtual EPP model; however, in FDTD numerical analysis, the spatial distribution of virtual EPPs, that is to say, the rate of the photochemical reaction, is represented by $I(r)\mathrm{d}I(r)/\mathrm{d}r$. Here, $I(r)$ is the photon intensity and $\mathrm{d}I(r)/\mathrm{d}r$ is the photon intensity gradient [3].

The model used in the FDTD simulation is shown in Fig. 4.33a. The first-stage nanorod is rectangular with a side of 100 nm, and the second-stage nanorod is a hexagonal rod with a side of 10 nm. The tip of each rod-like structure is pyramidal. A 100 nm portion at the tip of the second-stage nanorod is surrounded by SiO_2, corresponding to the silica gel film. These nanorods are periodically arrayed at a pitch of 20 nm. The light source was plane wave propagating to z- direction with 325 nm in wavelength. As optical constants, we assumed $n = 2.0$ and $k = 0.35$ for ZnO [55] and $n = 1.5$ and $k = 0$ for SiO_2 [56]. The calculated intensity distributions, $I(r)$, in the YZ plane ($x = 0$) and the XY plane ($z = 0$) are shown in Fig. 4.33b, c.

Curve X in Fig. 4.33d shows the spatial distribution of $I(r)$ at the SiO_2 surface as a function of the distance, d, from the ZnO nanorod surface. Here, the intensity was normalized to the value at $d = 0$ nm. The intensity was 0.45 at $d = 40$ nm and did not greatly decrease with increasing d. Therefore, when deposition of gold nanoparticles proceeds via the adiabatic process, gold nanoparticles should be deposited in the entire irradiated area, not just in the vicinity of the ZnO nanorods. In contrast, curve Y in Fig. 4.33d shows the distribution of $I(r)\mathrm{d}I(r)/\mathrm{d}r$, showing the distribution of virtual EPPs. The intensity gradient $\mathrm{d}I(r)/\mathrm{d}r$ is given by (4.9).

The values of $I(r)\mathrm{d}I(r)/\mathrm{d}r$ were also normalized to the value at $d = 0$ nm. Compared with $I(r)$, $I(r)\mathrm{d}I(r)/\mathrm{d}r$ showed a sharp decrease with increasing d. In the curves of $I(r)$ and $I(r)\mathrm{d}I(r)/\mathrm{d}r$, we compare the distances at which the intensity falls to $1/e$, as the attenuation length. $I(r)$ does not fall to $1/e$ at $I(50$ nm$)$, whereas the attenuation length for $I(r)\mathrm{d}I(r)/\mathrm{d}r$ is 0.9 nm. These results show that the effect of the nonadiabatic process is localized close to the nanorods. These simulation results show good agreement with the experimental result that the gold particles are in contact with the nanorod traces, indicating that the deposition reaction of gold nanoparticles deposited close to the ZnO nanorods in the

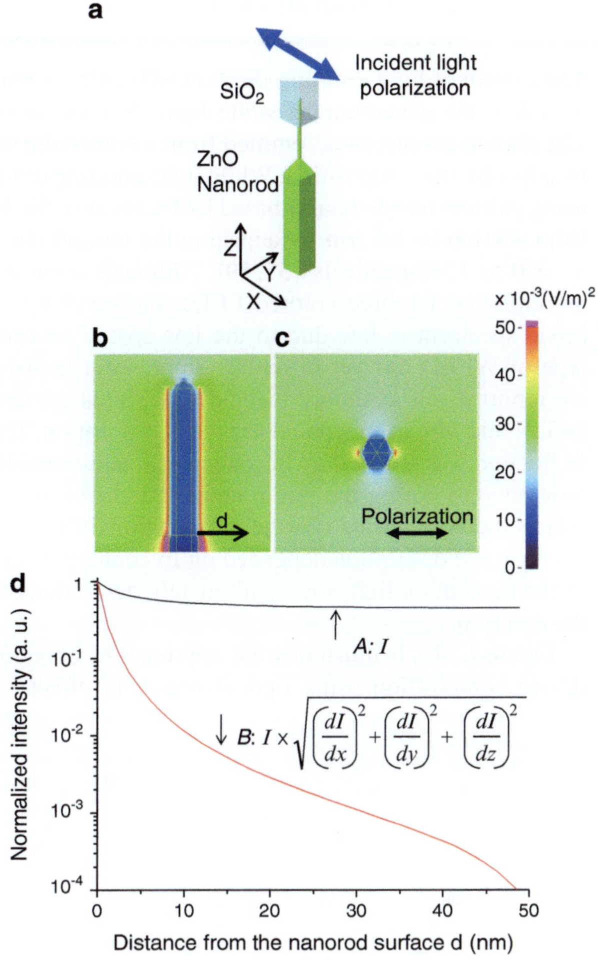

Fig. 4.33 (**a**) Configuration of model used in FDTD simulation. Calculated electric field intensity distributions in (**b**) YZ plane (axis center) and (**c**) XY plane (SiO₂ surface). (**d**) Intensity distributions versus distance from nanorod surface for I (curve A) and $I \times dI/dr$ (curve B)

experiment was due to the nonadiabatic process. In addition, as shown in Fig. 4.33c, the intensity distribution obtained from the FDTD calculation shows a strong polarization dependence. In the experiment, however, the gold nanoparticles were deposited in random directions relative to the nanorod traces. This difference arises because the deposition reaction of the gold particles is not an adiabatic reaction due to the excitation light but is a nonadiabatic reaction due to light emission from the ZnO nanorods. Because the exciton emission shows no polarization dependence, the nonadiabatic photoreaction which induces this exciton emission as a light source also shows no polarization dependence. Therefore, the directions of gold particle deposition are random.

4.4   Increased Spatial Homogeneity

The control of light-emitting-diode (LED) color is important in many applications, including the generation of white light [57] and in optical communications [58]. The photon energy, $h\nu_{em}$, emitted from a composite semiconductor can be tailored by adjusting its composition. White light-emitting diodes (WLEDs) were developed using gallium nitride (GaN)-based LEDs because the $h\nu_{em}$ from GaN can be shifted from 400 nm to 1.5 μm by adjusting the indium (In) content in $In_xGa_{1-x}N$ from $x = 0$ to 1, respectively [57, 59]. Although some commercial WLEDs combine the emission of three colored LEDs, the resulting color-rendering index over a broad spectrum is low due to the low spatial uniformity of In. As a result, this type of WLED has yet to replace fluorescent lamps in many applications. Here, we report a self-assembly method that yields greater spatial uniformity of In in InGaN thin film using optical near-field desorption. The spatial heterogeneity of the In fraction was reduced by introducing an additional light source (i.e., a desorption light source) during the photo-enhanced chemical vapor deposition (PECVD) of InGaN thereby causing near-field desorption of InGaN nanoparticles. The degree of nanoparticle desorption depended on In content of the film, and the photon energy of the desorption light source ultimately determined the emitted photon energy of the thin film.

Figure 4.34a, b illustrates the approach to obtaining a more spatially uniform device composition using optical near-field effects. During the initial stages of

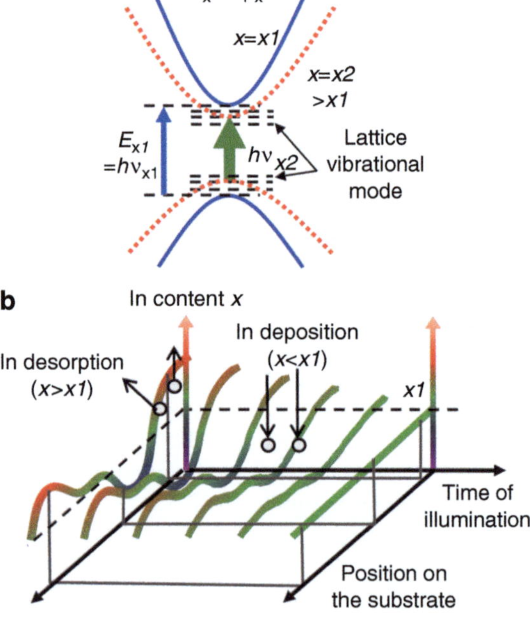

Fig. 4.34 Controlling the thin film composition using the optical near field. (a) A schematic diagram shows the energy dispersion and the generation of lattice vibrational modes induced by an optical near field on InGaN nanoparticles. (b) Spatial distributions of x are shown as a function of deposition time for In$_x$Ga$_{1-x}$N whilst illuminated with the desorption light source

$In_{x1}Ga_{1-x1}N$ nanoparticle growth (bandgap energy E_{x1}), a lattice vibrational mode can be excited by far-field light originating from an optical near field caused by coupling of exciton-polaritons and phonons (Fig. 4.34a) [3]. If the nanoparticles are concurrently illuminated by a desorption source with photon energy of $h\nu_{x2}(< E_{x1})$, a strong optical absorption due to a multistep excitation of lattice vibrational modes induces desorption of a fraction of the nanoaparticle population [21]. The absorption is enhanced by increasing the In content. As the deposition proceeds with desorption source illumination, the growth is governed by a trade-off between In deposition (where the In content $x < x_1$) and In desorption (where $x > x_1$). Thus, the resulting In content in the film is a function of $h\nu_{x2}$, and both spatial heterogeneity of the In fraction and spectral broadening (Fig. 4.34b) are avoided.

A spectral change was observed upon introduction of the desorption light source during PECVD of InGaN at room temperature [60] in which $h\nu_{em}$ was determined by the photon energy $h\nu_{x2}$. A 5th-harmonic, Q-switched Nd:YAG laser ($h\nu_{depo} = 5.82\,eV$, $\lambda = 213\,nm$) was used to excite and photodissociate trimethylgallium (TMG), triethylindium (TEI), and ammonia (NH_3, 99.999%). The choice of laser was based on the strong photo-absorptions of gas-phase TMG ($E_g > 4.59\,eV$), TEI ($E_g > 4.77\,eV$) and NH_3 ($E_g > 5.66\,eV$), respectively [61, 62]. The desorption light source was introduced through an optical window, and H_2 gas was introduced around the window to prevent GaN deposition onto the window. The substrate was placed at the center of the reaction chamber and irradiated with a 2-mm spot size of excitation light. The total pressure in the reaction chamber was 5.4 Torr, and the deposition time was 60 min.

The morphology of the GaN sample was investigated with a scanning electron microscope (SEM). Figure 4.35a shows the overall SEM image of the InGaN film on the substrate. Figure 4.35b–d shows magnified SEM images from within the white dashed circle in Fig. 4.35a. Similar morphologies, consisting of 100-nm lines, were observed at different TEI flow rates, r_{TEI}. The relative atomic compositions of indium, gallium, and nitrogen were obtained by monitoring photoluminescence (PL) induced by a continuous wave He–Cd laser (3.81 eV, $\lambda = 325\,nm$), in which the PL peak was shifted by changing r_{TEI} (Fig. 4.35a). The In content, x, was determined according to

$$E = 3.42 - 4.95x \qquad (4.10)$$

where E refers to the PL peak energy [63]. Figure 4.36b shows that x was a linear function of r_{TEI}, which agrees with results obtained using low-temperature metal-organic chemical vapor deposition (MOCVD) [64] (Fig. 4.36).

Based on the above results, a spectral shift was induced by introducing desorption light during PECVD. InGaN was grown with $r_{TEI} = 2.5 \times 10^{-3}$ sccm and under illumination with a desorption light source of $h\nu_{x2} = 2.71$ eV. As shown in Fig. 4.37a, the PL intensity in the region $h\nu_{em} < h\nu_{x2}$ (curve X′) decreased relative to that observed in the absence of desorption light (curve X). The difference in PL intensity, I_{diff}, between X and X′ (Fig. 4.37b) clearly shows the decrease in the PL intensity of X′ at energies less than $h\nu_{x2}$, indicating near-field desorption as described in Fig. 4.34a.

Fig. 4.35 SEM images of fabricated InGaN thin films. The overall image is shown in (**a**). Magnified images show films fabricated with $r_{\text{TEI}} = 0$ (**b**), 2.5×10^{-3} (**c**), 5.0×10^{-3} (**d**) sccm

Fig. 4.36 Spectral control by adjusting the TEI flow rate, r_{TEI}, at room temperature. (**a**) PL spectra obtained at 5 K were acquired following room-temperature film deposition with $r_{\text{TEI}} = 0$ (α), 2.5×10^{-3} (β), 5.0×10^{-3} (γ) sccm with $r_{\text{TMG}} = 0.5$ sccm. (**b**) The indium content of InGaN is shown as a function, r_{TEI}

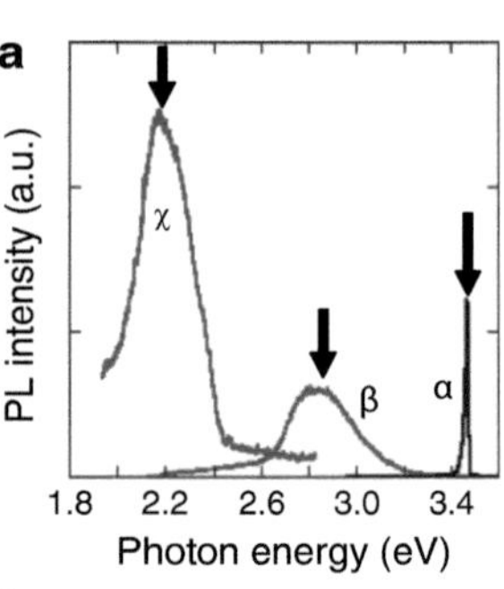

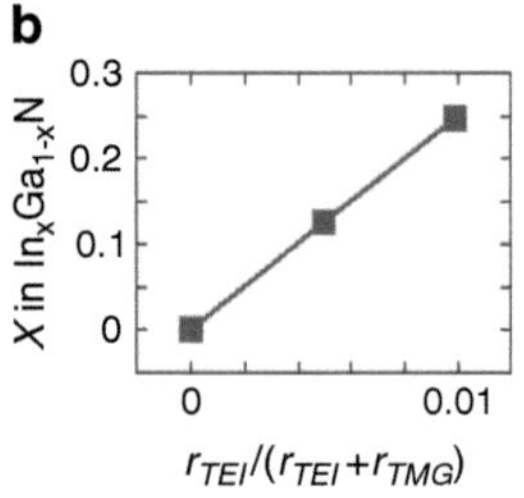

Fig. 4.37 Spectral changes induced by desorption light illumination at 2.71 eV during PECVD. (**a**) PL spectra were obtained at 5 K using room-temperature PECVD with $r_{TEI} = 2.5 \times 10^{-3}$ (curves X and X') and a desorption source energy of $h\nu_{x2} = 2.71$ eV. (**b**) The difference spectrum is shown for X–X'

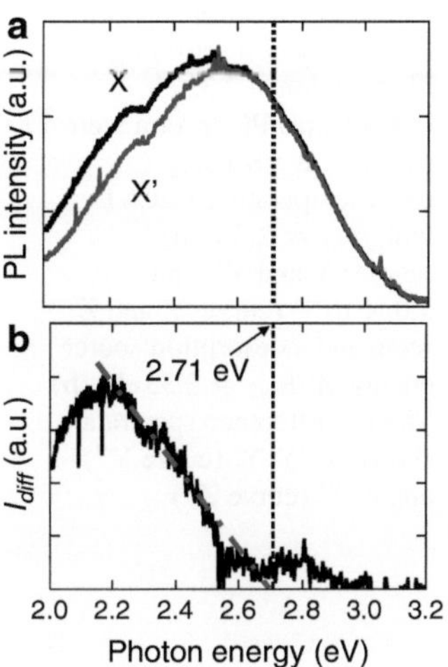

Using a desorption light source with $h\nu_{x2} = 2.33$ eV, which is lower in energy than the peak PL of deposited InGaN (2.5 eV) in the absence of desorption light, similar decreases in PL intensity were observed for $r_{TEI} = 2.5 \times 10^{-3}$ sccm and $r_{TEI} = 5.0 \times 10^{-3}$ sccm (curves Y' and Z' in Fig. 4.38a, respectively). The difference in PL between Y and Y', and Z and Z', as indicated by curves Y'' and Z'' in Fig. 4.38b, respectively, shows that the PL intensity of both Y' and Z' decreased at energies below $h\nu_{x2} = 2.33$ eV. In addition, Fig. 4.38b shows that higher levels of r_{TEI} (curve Z'') resulted in increased I_{diff} values at $h\nu_{x2} > 2.33$ eV relative to those obtained at lower r_{TEI} (curve Y''). This further indicates In desorption. This result confirms that the use of the desorption light source during film deposition did not permit additional In doping other than that determined by the photon energy of the desorption light source itself. This effect resulted in a film with a more spatially uniform In content.

4.5 Improving the Device Efficiency Using the Phonon-Assisted Process

The phonon-assisted process can be utilized to the fabrication of the device with an optimum structure for inducing the phonon-assisted process when they are utilized in operation. This fabrication process was applied to fabricate (3-hexylthiophene)(P3HT) and ZnO as test specimens for the p-type and n-type semiconductors, respectively. A poly P3HT was used as a test specimen of a p-type semiconductor material for the transducer device because it was straightforward to fabricate its thin film without using any special professional tools. Preliminary experiments confirmed that (1) the used P3HT had the maximum photo-absorption

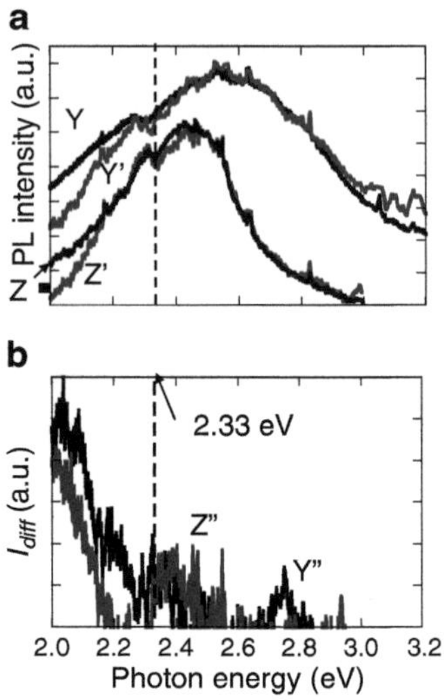

Fig. 4.38 Spectral changes induced by desorption light illumination at 2.33 eV during PECVD. (**a**) PL spectra were obtained at 5 K using room-temperature PECVD with $r_{\text{TEI}} = 2.5 \times 10^{-3}$ (curves Y and Y′) and 5.0×10^{-3} (curves Z and Z′) sccm and a desorption source energy of $h\nu_{x2} = 2.33$ eV. (**b**) The PL difference spectra are shown for Y–Y′ (curve Y″) and Z–Z′ (curve Z″)

at the wavelength $\lambda_{\text{p}} = 430$ nm, (2) its long-wavelength cutoff λ_{c} was 570 nm, and (3) the magnitude of the photo-absorption at λ_{c} was less than 1/100 times that at λ_{p} which was negligibly low for the present study. A ZnO thin film was used as a n-type semiconductor material because it was transparent in the range of the wavelength longer than 400 nm. Further, its valence and conduction bands are located so that they did not prevent the photocurrent generation. The ITO and Ag films were used as two electrodes. The P3HT plays a main role for the device operation because the depletion layer of the pn-junction are formed in the P3HT.

As is schematically explained by Fig. 4.39a, thin films of these materials were deposited on a sapphire substrate by the series of the processes, which are:

1. After the successive ultrasonic cleanings of the sapphire substrate using acetone, methanol, and pure water, respectively, the ITO film of 200 nm thickness was deposited on the sapphire substrate by the sputtering.
2. The ZnO film of 100-nm thickness was deposited on the ITO film by the sputtering.
3. After the chloroform solution of the P3HT of 10 mg/ml was spin-coated on the ZnO film, it was baked up to 120°C for 6 h in vacuum in order to remove impurities such as the residual chloroform. As a result, the P3HT film of 50 nm thickness was formed.
4. An Ag thin film of a few nanometers was deposited on the P3HT film. As a result, the multilayered films of 30-mm²-area was obtained on the sapphire substrate.

The last stage of the fabrication process further deposited Ag on the Ag thin film of (4) in order to induce the phonon-induced process for efficient photocurrent generation. In order to develop a novel method for further deposition of Ag, the authors obtained a crucial insight from their previous works on a size- and

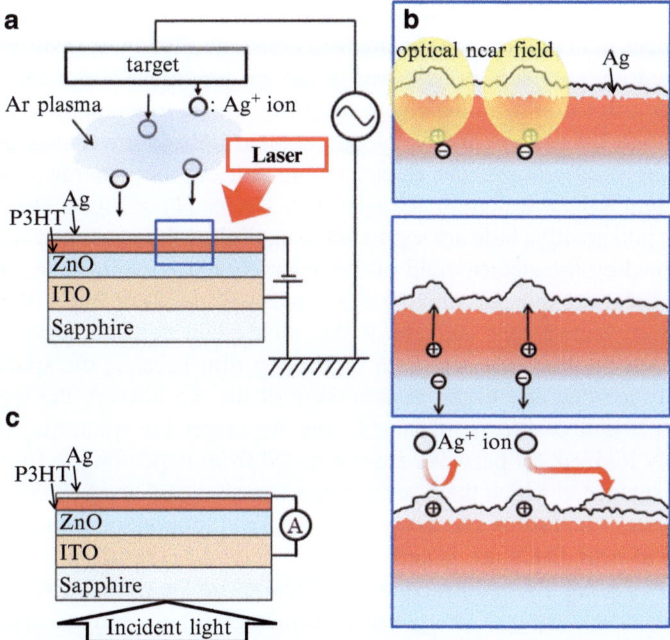

Fig. 4.39 Schematic explanation of fabricating a photoelectric transducer device. (**a**) Cross-sectional structure of the films stacked on a sapphire substrate. Thickness of respective film Ag: a few nm, P3HT: 50 nm, ZnO: 100 nm, ITO: 200 nm. (**b**) Schematic explanation of the Ag deposition. In the figures (1)–(3), the $+$ and $-$ in the *circles* represent the positive hole and electron, respectively. (**c**) Application of the incident light from the rear surface of the sapphire substrate for photocurrent detection

position-controlled self-assembling optical near field deposition of nanometric metallic particles [21]. As is schematically explained by Fig. 4.39b, the Ag was deposited by the RF-sputtering under light illumination on the previously deposited Ag thin film of (4) while the pn-junction of P3HT/ZnO was reversely biased with the D.C. voltage V_b. Here, the V_b was fixed to -1.5 V, and the wavelength λ_0 of the incident light was 660 nm, which is longer than λ_c of the P3HT. It is expected that the deposition forms the unique granular surface of the Ag film originated from the phonon-assisted process induced by the optical near field due to light illumination and reverse bias. A notable phonon-assisted process originated from this granular surface is also expected for photocurrent generation.

The origins of the unique granular surface are:

1. Under light illumination, the optical near fields are generated on the Ag surface, which is deposited by the RF-sputtering in a vacuum chamber filled with the argon gas of 0.6 Pa-pressure. This optical near field excites the coherent phonon at the pn-junction and resulting the exciton-phonon-polarition generation, which is the coupled state of the optical near field and coherent phonon [3, 65, 66].

Since the energy of the exciton-phonon-polarition is the sum of those of the optical near field photon and the induced coherent phonon, it is sufficiently high for generating an electron–hole pair at the pn-junction via the electric dipole-forbidden transition from the electronic ground level to the phonon energy level even though the photon energy of the incident light is lower than E_g. By this phonon-assisted process, the electron–hole pair can be generated even though the incident light wavelength is longer than λ_c (see (1) of Fig. 4.39b).

2. Electron and positive hole are separated to annihilate the generated electron–hole pair, forced by the electric field of the reversely biased voltage. As a result, the positive holes are injected into the deposited Ag (see (2) of Fig. 4.39b).

3. It has been known that the rate of the subsequent deposition of Ag strongly depends on the local electric field on the Ag film because the sputtered Ag is positively ionized due to the transmission of the Ag through the argon plasma or the collision of the Ar-plasma to the Ag-target for sputtering [67]. These positively ionized Ag particles are prevented from depositing to the area of the Ag film surface in which the positive holes have been injected by the optical near field. It means that subsequent deposition of Ag is suppressed in the area in which the optical near fields are generated effectively. As a result, a unique granular surface of the Ag film is formed, which depends on the spatial distribution of the optical near field energy. This granular surface grows in a self-assembled manner with increasing the RF-sputtering time (see (3) of Fig. 4.39b).

By using the Ag film with the unique granular Ag film described above as an electrode of the photoelectric transducer device and by applying the incident light from the rear surface of the sapphire substrate (Fig. 4.39c), it is expected that the optical near field can be effectively generated on the electrode. Thus, electron–hole pairs can be generated effectively by the phonon-assisted process if it is illuminated by the light with the same wavelength λ_0 as the one used for the above mentioned deposition by the RF-sputtering. On the other hand, if it is illuminated by the light the different wavelength λ_1, spatial profile and the photon energy of the generated optical near field are different from those of the case of λ_0. Therefore, efficiency of the electron–hole pair generation should be lower. Thus, this device should exhibit the wavelength selectivity in the photocurrent generation efficiency, which is the highest at the wavelength λ_0. Furthermore, since this wavelength is longer than λ_c, the working wavelength becomes longer than the one beyond E_g.

By employing several values of V_b and the incident light power P, Ag was deposited on the previously fabricated Ag thin film of (4) by the RF-sputtering with the sputtering time of 30 min under a reverse bias and light illumination. Finally, five devices were fabricated, for which the values of V_b and P were $V_b = 0$ and $P0$ (device 1), $V_b = 0$ and $P30$mW (device 2), $V_b = 0$ and $P80$mW (device 3), $V_b = -1.5$ V and $P50$mW (device 4), and $V_b = -1.5$ V and $P70$mW (device 5). It should be noted that the Ag was deposited for the device 1 by the conventional RF-sputtering without applying V_b and P. This device was used as the reference to evaluate the performances of other devices.

Figure 4.40a–e represents the SEM images of the Ag-film surfaces of the devices 1–5. By comparing them, it is easily found that the Ag surfaces of the devices 4 and 5 in Fig. 4.40d, e are very rough with larger grains than those of the devices 1–3 (Fig. 4.40a–c), and thus, their thicknesses are spatially inhomogenous. Furthermore, the grain sizes in Fig. 4.40d, e are different between each other, which is originated from the phonon-assisted process utilized to control the morphology by the self-organized manner, as was described by (1)–(3). Figure 4.40d, e were analyzed to evaluate the size distribution of the grains. These figures were digitized and the shapes of the grains were approximated by circles of equal area. After this approximation, the distribution of the diameters were fitted by a lognormal size-distribution function, which has been popularly used for representing the size

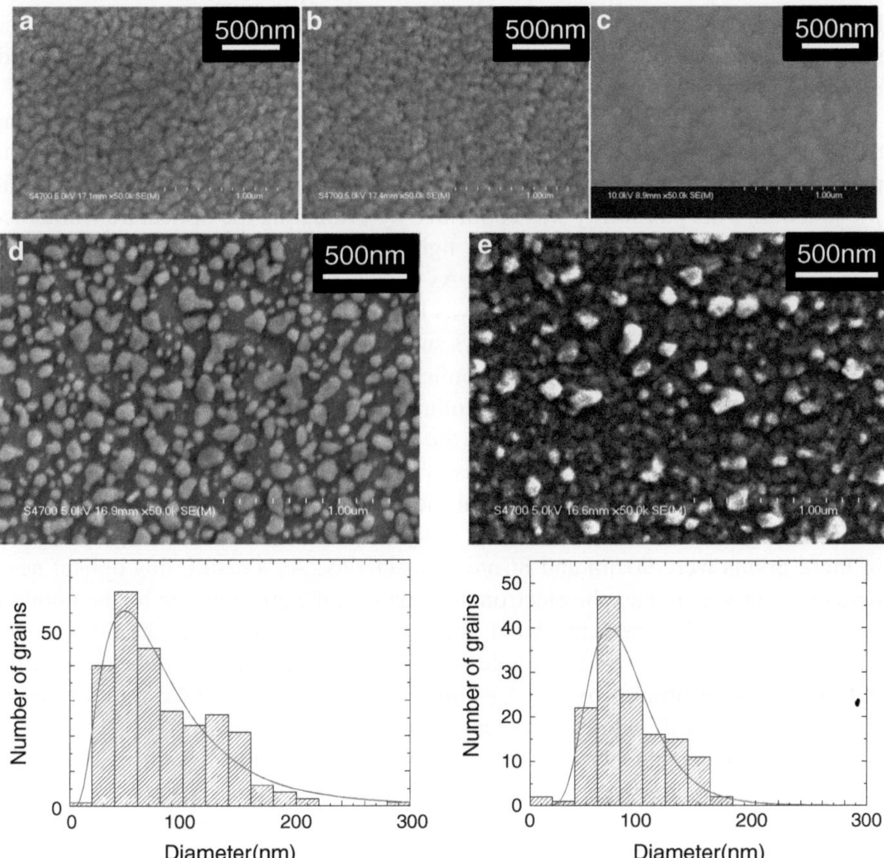

Fig. 4.40 SEM images of the Ag-film surfaces. **(a)**–**(e)** is for the devices 1–5, respectively. Histograms in **(d)** and **(e)** represent the distributions of the diameters of the grains in the SEM image. The solid curves in these figures represent the lognormal size-distribution fitted to these histograms

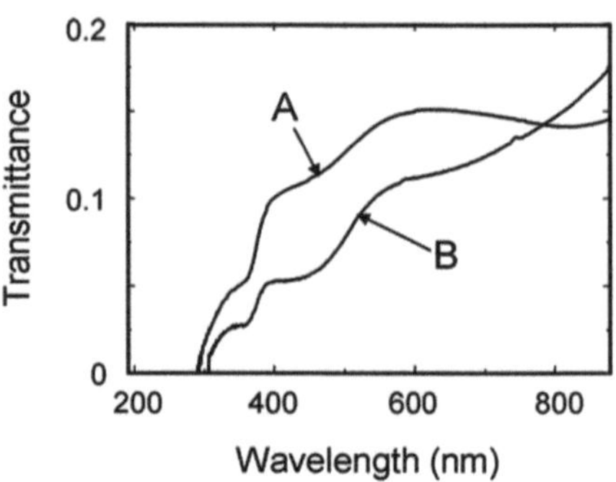

Fig. 4.41 Optical transmittance of the device measured by applying the incident light from the rear surface of the sapphire substrate. The curves A and B are for the device 5 and 4, respectively

distribution in particle growth process [(1) of [68]]. The results are shown by Fig. 4.40d, e. From the fitted lognormal size-distribution curves, the average and standard deviation of the diameters were found to be 90 nm and 64 nm in Fig. 4.40d, respectively. Those of Fig. 4.40e were 86 nm and 32 nm, respectively.

Since the decay lengths of the optical near-field on a nanometric particle along the directions in parallel and normal to its substrate surface are equivalent to the particle size, it is possible for the optical near-field on these grains to reach the pn-junction. This possibility was confirmed by evaluating the optical transmittance of the devices by applying the incident light (light beam diameter: 2 mm) from the rear surface of the sapphire substrate. As shown by the curve A of Fig. 4.41, the optical transmittance of the device 5 was higher than 0.1 in the wavelength range of 400–880 nm. The curve B also shows such the high optical transmittance for the device 4. From these curves, it was confirmed that the thicknesses of the Ag film of Fig. 4.40d, e averaged over the illuminated area, on which a high aspect-ratio grains were formed, were thinner than the Agfs optical penetration depth of about 10 nm. Thus, the sum of the thicknesses of the Ag and P3HT was estimated as thin as 60 nm. Therefore, it is expected that the optical near-field generated on the Ag grains of Fig. 4.40d, e can reach to the pn-junction because the average diameters of these grains were 90 nm and 86 nm, respectively. As a result, this optical near-field efficiently generates the electron–hole pairs at the pn-junction by the phonon-assisted process. On the other hand, the optical transmittance of the devices 1–3 were as low as 1×10^{-5} in the whole wavelength range of Fig. 4.41 because of the reflection and absorption by the Ag films. Their thicknesses were found to be as thick as 800 nm by direct measurements using a surface profiler.

Wavelength dependences of the photocurrent generations were evaluated by using a Ti:Al$_2$O$_3$ laser-pumped wavelength-tunable OPO as a light source. The light beam diameter was 1mm. Experimental results for the wavelength range of 580 nm $\leq \lambda_i \leq$ 670 nm are shown by Fig. 4.42, in which the incident light power was fixed to 1 mW. The relation between the generated photocurrent and the incident light power was linear. The photocurrents from the devices 2 and 3 are not shown in this figure because they were negligibly low. That from the device 1 was also negligible, but it is shown by the curve A as a reference. The curves B and C present

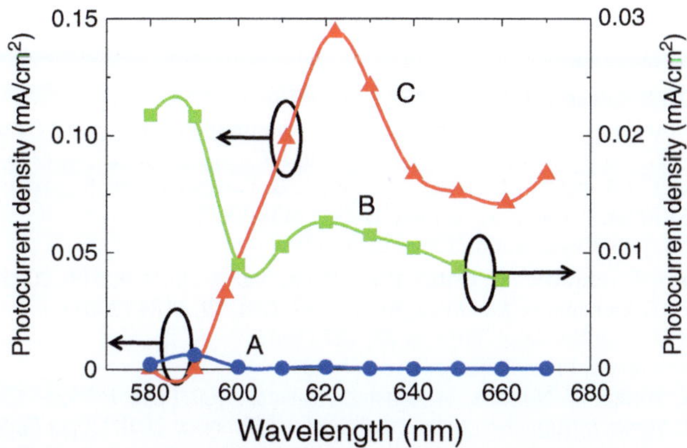

Fig. 4.42 Dependences of the generated photocurrent and the wavelength of the incident light. The curves A, B, and C are for the devices 1, 4, and 5, respectively

significant increases of the photocurrents from the devices 4 and 5, respectively, up to λ_i 670 nm, which clearly confirms the expansion of the working wavelength range of the photoelectric transducer beyond the λ_c. These curves represent the possibility of photocurrent generation even in the wavelength range over 670 nm.

The curve C shows that the photocurrent from the device 5 is the highest at $\lambda_{ih} = 620$ nm while it is very low in the range of $\lambda_i \leq 600$ nm, which is due to the conspicuous phonon-assisted process. Thus, the device 5 can be nothing but a wavelength-selective photoelectric transducer for the incident light of the wavelength beyond λ_c to be proposed here. It should be noted that this wavelength λ_{ih} (=620 nm) is 40 nm shorter than λ_0 (=660 nm) used for depositing Ag by the RF-sputtering. This difference can be attributed to the D.C. Stark effect induced by the reversely biased voltage V_b(=− 1.5 V) applied at the process of Ag deposition: It has been known that the width of the depletion layer of the pn-junction is about 10 nm for general organic semiconductors [69]. Also, the relative dielectric constant of the P3HT is 3.0 [70]. By using these values, it is estimated that the D.C. electric field of 1×10^6 V/m was applied to the depletion layer of the P3HT/ZnO due to V_b. Furthermore, by approximating the reduced mass of the electron–hole pair in the pn-junction as the mass of the electron in vacuum, the shift of the long-wavelength cutoff λ_c induced by the applied D.C. electric field was estimated by using a formula of the photo-absorption coefficient ((23) of reference [71]). The estimated value was 40 nm, which agrees with the measured difference given above.

The curve B of Fig. 4.42 also represents that the device 4 generated the photocurrent even in the range of 600 nm, which means that it has a wider working wavelength range and can be advantageously used for, e.g., a solar cell battery device.

References

1. Y. Tanaka, K. Kobayashi, J. Micros. **229**, 228 (2008)
2. A. Sato, Y. Tanaka, F. Minami, K. Kobayashi, J. Luminescence **129**, 1718 (2009)
3. T. Kawazoe, K. Kobayashi, S. Takubo, M. Ohtsu, J. Chem. Phys. **122**, 024715 (2005)
4. T. Kawazoe, T. Takahashi, M. Ohtsu, Appl. Phys. B: Lasers Opt. **98**, 5 (2010)
5. B. Wua, A. Kumar, J. Vac. Sci. Technol. B **25**, 1743 (2007)
6. L.M. Cook, J. Non-Cryst. Solids **120**, 152 (1990)
7. K. Kobayashi, T. Kawazoe, M. Ohtsu, IEEE Trans. Nanotechnol. **4**, 517 (2005)
8. T. Kawazoe, Y. Yamamoto, M. Ohtsu, Appl. Phys. Lett. **79**, 1184 (2001)
9. R. Kullmer, D. Bäuerle, Appl. Phys. A **43**, 227 (1987)
10. T. Izawa, N. Inagaki, Proc. IEEE **68**, 1184 (1980)
11. T. Yatsui, K. Hirata, W. Nomura, M. Ohtsu, Y. Tabata, Appl. Phys. B **93**, 55 (2008)
12. P. Stoica, R. Moses *Introduction to Spectral Analysis* (Prentice-Hall, Upper Saddle River, 1997)
13. M. Naruse, T. Yatsui, W. Nomura, K. Hirata, Y. Tabata, M. Ohtsu, J. Appl. Phys. **105**, 063516 (2009)
14. F. Ladouceur: J. Light. Technol. **15**, 1020 (1997)
15. L.H. Koopmans *The Spectral Analysis of Time Series* (Academic, New York, 1974)
16. Y. Akita, Y. Kato, M. Hosaka, Y. Ono, S. Suzuki, A. Nakajima, M. Yoshimoto, Mater. Sci. Eng. B **161**, 151 (2009)
17. L.B. Glebov, E.N. Boulos, J. Non-Cryst. Solids **242**, 49 (1998)
18. L. Skuja, J. Non-Cryst. Solids **239**, 16 (1998)
19. F.J. Himpsel, J.E. Ortega, G.J. Mankey, R.F. Willis, Adv. Phys. **47**, 511 (1998)
20. E.J. Menke, Q. Li, R.M. Penner, Nano Lett. **4**, 2009 (2004)
21. T. Yatsui, W. Nomura, M. Ohtsu, Nano Lett. **5**, 2548 (2005)
22. A product of the Covalent Materials Corporation, http://www.covalent.co.jp/
23. J.B. Wachtman, A.R. Haber *Ceramic Films and Coatings* (William Andrew Publishing, Noyes, New York, Norwich, 1993)
24. F. Benabid, M. Notcutt, V. Loriette, L. Ju, D.G. Blair, J. Phys. D **33**, 589 (2000)
25. R.O. Duda, P.E. Hart, Commun. ACM **15**, 11 (1972)
26. M. Ohtsu, T. Kawazoe, T. Yatsui, M. Naruse, IEEE J. Selec. Top. Quant. Electron. **14**, 1404 (2008)
27. T. Yatsui, S. Sangu, T. Kawazoe, S.J. An, J. Yoo, G.-C. Yi, Appl. Phys. Lett. **90**, 223110 (2007)
28. T. Kawazoe, K. Kobayashi, K. Akahane, M. Naruse, N. Yamamoto, M. Ohtsu, Appl. Phys. B **84**, 243 (2006)
29. T. Kawazoe, K. Kobayashi, M. Ohtsu, Appl. Phys. Lett. **86**, 103102 (2005)
30. T. Ishikawa, S. Kohmoto, K. Asakawa, Appl. Phys. Lett. **73**, 1712 (1998)
31. R.D. Piner, J. Zhu, F. Xu, S. Hong, C.A. Mirkin, Science **283**, 661 (1999)
32. H. Koyama, N. Koshida, J. Appl. Phys. **74**, 6365 (1993)
33. K. Kitamura, T. Yatsui, M. Ohtsu, Opt. Rev. **13**, 222 (2006)
34. T. Yatsui, M. Kourogi, K. Tsutsui, J. Takahashi, M. Ohtsu, Opt. Lett. **25**, 1279 (2000)
35. T. Yatsui, K. Itsumi, M. Kourogi, M. Ohtsu, Appl. Phys. Lett. **80**, 2257 (2002)
36. N. Yamamoto, H. Takagi, Thin Solid Films **388**, 138 (2001)
37. A.G. Cullis, L.T. Canham, Nature **353**, 335 (1991)
38. T. Kawazoe, M. Ohtsu, Y. Inao, R. Kuroda, J. Nanophot. **1**, 011595 (2007)
39. E.D. Palik *Handbook of Optical Constants of Solids* (Academic press, New York, 1985)
40. E. Hutter, J.H. Fendler, Adv. Mater. **16**, 1685 (2004)
41. W. Nomura, T. Yatsui, M. Ohtsu, Appl. Phys. B **84**, 257 (2006)
42. K.I. Bolotin, F. Kuemmeth, A.N. Pasupathy, D.C. Ralph, Appl. Phys. Lett. **84**, 3154 (2004)
43. W. Nomura, M. Ohtsu, T. Yatsui, Appl. Phys. Lett. **86**, 181108 (2005)
44. M.-C. Daniel, D. Astruc, Chem. Rev. **104**, 293 (2004)
45. I. Tanahashi, Y. Manabe, T. Tohda, S. Sasaki, A. Nakamura, J. Appl. Phys. **79**, 1244 (1996)

46. I. Tanahashi, T. Tohda, J. Am. Ceram. Soc. **79**, 796 (1996)
47. M. Sugiyama, S. Inasawa, S. Koda, T. Hirose, T. Yonekawa, T. Omatsu, A. Takami, Appl. Phys. Lett. **79**, 1528 (2001)
48. T. Yatsui, M. Ohtsu, Appl. Phys. Lett. **95**, 043104 (2009)
49. K. Kitamura, T. Yatsui, M. Ohtsu, G.-C. Yi, Nanotechnology **19**, 175305 (2009)
50. W.I. Park, J. Yoo, G.-C. Yi, J. Korean Phy. Soc. **46**, L1067 (2005)
51. A. Wander, N.M. Harrison, Surf. Sci. **457**, L342 (2000)
52. A. Wander, N.M. Harrison, Surf. Sci. **468**, L851 (2000)
53. B. Meyer, D. Marx, Phys. Rev. B **67**, 035403 (2003)
54. D.C. Reynolds, D.C. Look, B. Jogai, C.W. Litton, T.C. Collins, W. Harsch, G. Cantwell, Phys. Rev. B **57**, 12151 (1998)
55. T.W. Kim, K.D. Kwack, H.-K. Kim, Y.S. Yoon, J.H. Bahang, H.L. Park, Solid State Commun. **127**, 635 (2003)
56. R. Thielsch, A. Gatto, J. Heber, N. Kaiser, Thin Solid Films **410**, 86 (2002)
57. S. Nakamura, Science **281**, 956 (1998)
58. K.J. Vahala, Nature **424**, 839 (2003)
59. S. Nizamoglu, T. Ozel, E. Sari, H.V. Demir, Nanotechnology **18**, 065709 (2007)
60. S. Yamazaki, T. Yatsui, M. Ohtsu, Appl. Phys. Exp. **1**, 061102 (2008)
61. H. Okabe, M.K. Emadi-Babaki, V.R. McCrary, J. Appl. Phys. **69**, 1730 (1991)
62. K. Watanabe, J. Chem. Phys. **22**, 1564 (1954)
63. M.D. McCluskey, C.G. Van de Walle, C.P. Master, L.T. Romano, N.M. Johnson, Appl. Phys. Lett. **72**, 2725 (1998)
64. A. Koukitu, H. Seki, Jpn. J. Appl. Phys. **35**, L1638 (1996)
65. K. Kobayashi, S. Sangu, H. Ito, M. Ohtsu, Phys. Rev. A **63**, 013806 (2000)
66. Y. Tanaka, K. Kobayashi, Phys. E **40**, 297 (2007)
67. J. Jooi, J. Vac. Sci. Technol. A **18**, 23 (2000)
68. J. Soederlund, L.B. Kiss, G.A. Niklasson, C.G. Granqvist, Phys. Rev. Lett. **80**, 2386 (1998)
69. M. Onoda, K. Tada, H. Nakayama, J. Appl. Phys. **86**, 2110 (1999)
70. A. Khaliq, F. Xue, K. Varahramyan, Microelectron. Eng. **86**, 2312 (2009)
71. J. Callaway, Phys. Rev. **130**, 549 (1963)

Chapter 5
Some Remarks and Outlook

5.1 Remarks

Previous chapters described how the dressed-photon and phonon are used. Additional achievements can be realized in various applications. Recent progress from new trials is reviewed, and some remarks are presented in following chapter.

5.1.1 Photolithography

The phonon-assisted process can be applied to photolithography and used to pattern widely available commercial photoresists using a visible light source even though such photoresists are sensitive only to UV light. Because the size of localized dressed photon and phonon is extremely small, a 22-nm half pitch pattern can be fabricated [1]. Other examples of fabricated structures include diffraction gratings and Fresnel zone plates for soft X-rays with a wavelength of 0.5–1.0 nm [2]. It should be pointed out that these devices are fabricated by using green light with a wavelength more than 500 times longer than that of soft X-rays. Advanced lithography uses an EUV light source with a wavelength as small as 13.5 nm. However, the EUV light source has extremely high power consumption.

Visible light lithography offers a new approach for advanced lithography. In addition to the high resolution obtained using a visible light source, this method, based on the phonon-assisted process, has several advantages.

1. *Multiple exposure*: Because a visible light source exposes only the area where the dressed photon and phonon are generated, other areas can be utilized.
2. *Patterning of optically inactive films*: Similar to the dissociation of optically inactive $Zn(acac)_2$ molecules by phonon-assisted PCVD, this method can pattern even optically inactive films, such as a ZEP-520A, which is popularly used as a resist film for EB lithography [2]. Because the resist for EB lithography has a lower molecular mass compared with that for photolithography, a further decrease in resolution is expected.

T. Yatsui, *Nanophotonic Fabrication*, Nano-Optics and Nanophotonics,
DOI 10.1007/978-3-642-24172-7_5, © Springer-Verlag Berlin Heidelberg 2012

5.1.2 Near-Field Etching

Because near-field etching is an optical chemical reaction, it is applicable to a variety of substrates, including amorphous (glass, plastic, and so on) and crystal (diamond, SiC, GaN, Si, and so on) substrates. This technique is a noncontact method and does not require a polishing pad; thus, it can be applied not only to flat substrates but also to three-dimensional substrates that have convex or concave surfaces, such as micro-lenses and the inner wall surface of cylinders. Furthermore, this method is also suited for with mass-production. The near-field etching technique does not require CeO_2, which is necessary for conventional chemical mechanical polishing. By optimizing the etching conditions, we developed a polishing method that does not require a rare metal.

5.1.3 Light Emitting Devices

Using phonon-assisted fabrication, indirect band gap semiconductors exhibit bright emission [3]. Using stimulated emission triggered by an optical near field generated at the inhomogeneous domain boundary of B-doped Si, T. Kawazoe et al. controlled annealing with current injection and fabricated a high-efficiency, broadband light-emitting diode (LED). This device emitted light via a two-step phonon-assisted process due to the optical near field generated at the inhomogeneous domain boundary of B. Even though they used bulk crystal Si with a simple homojunction structure, the emission band of the device extended over energies of 0.73–1.24 eV with 11 W of input electrical power, and the total optical power was as high as 1.1 W. The external power conversion efficiency was 1.3%, the differential external power conversion efficiency was 5.0%, the external quantum efficiency was 15%, and the differential external quantum efficiency was 40%.

5.1.4 Nanophotonic Energy Conversion

5.1.4.1 Optical/Optical Energy Up-conversion

In order to confirm the optical energy up-conversion, powdery grains of DCM organic dye molecules were placed in a quartz container and used as a test material [4]. Although the absorption band-edge wavelength of the DCM is as short as 670 nm, the grains were illuminated by near infrared light of 805 nm wavelength. This illumination generated the dressed-photon and phonon at the edges of grains, which were then exchanged between the adjacent grains. As a result of this exchange, an electron in the adjacent grain was excited by a phonon-assisted process, and then, it emitted light whose photon energy was higher than that of the incident light due to the contribution of the phonon energy. The efficiency of the frequency up-conversion was confirmed to be much higher than that of the efficiency of conventional second harmonic generation from a popular KDP crystal, which has the same optical thickness as that of the powdery grains of DCM organic dye molecules placed in the quartz container. The efficiency of the frequency up-conversion was more than 100 times higher than that of the second harmonic generation for an incident light power density lower than 1 W/cm^2.

5.1.4.2 Hydrogen Generation

Because conventional photocatalysts are activated only under UV light irradiation, the effective use of visible light is a very important goal. Hydrogen generation with visible light using a phonon-assisted optical near-field process can be realized. Carrier excitation using propagating light requires photon energy higher than the band gap energy (3.3 eV for ZnO). An optical near field can excite coherent phonons in nanoscale structures, i.e., a phonon-assisted process. To realize an efficient phonon-assisted process, we introduced ZnO nanorods grown by metal-organic vapor-phase epitaxy (MOVPE) as electrodes. The optical near field generated around a material depends on the material size. In the MOVPE process, the temperature controls the diameter of the ZnO nanorods. Water was electrolyzed using the electrodes, and hydrogen was generated at the Pt counter electrode. The current was measured under visible light irradiation (2.6 eV). The current of the sample with 10-nm nanorods was more than 20 times greater than that of the sample with 100-nm nanorods. Additionally, the current of the sample with 10-nm nanorods was more than 70 times greater than that of the bulk single crystal ZnO substrate with a flat surface. Photoluminescence measurements revealed that visible light excitation did not originate from the impurities. Visible light excitation occurred due to a phonon-assisted optical near field generated by introducing ultrafine ZnO nanorod structures as small as 10 nm. Further increases in the efficiency of hydrogen can be realized using photocatalysts fabricated using the dressed-photon and phonon assisted process, as described in the previous section.

5.2 Summary

The study of nanophotonics has revealed that novel functions can be emerged in devices when they are fabricated using dressed-photon and phonon. Because dressed-photon and phonon has higher energy than incident light, their use can promote effective use of infrared light (heat). Thus, dressed-photon and phonon will be indispensable for future sustainable development.

References

1. Y. Inao, S. Nakasato, R. Kuroda, M. Ohtsu, Sci. Direct Microelectron. Eng. **84**, 705 (2007)
2. M. Ohtsu (ed.) *Progress in Nanophotonics* (Springer, Berlin, 2011)
3. T. Kawazoe, M.A. Mueed, M. Ohtsu, Appl. Phys. B: Lasers Opt., **104**, 747 (2011)
4. T. Kawazoe, H. Fujiwara, K. Kobayashi, M. Ohtsu, IEEE J. Select. Topics Quant. Electron. **15**, 1380 (2009)
5. K. Kitamura, T. Yatsui, H. Yasuda, T. Kawazoe, M. Ohtsu, *E-MRS ICAM IUMRS Spring Meeting 2011*, Nice, France (paper number: 14-8), 9–12 May 2011

Index

T. Yatsui, *Nanophotonic Fabrication*, Nano-Optics and Nanophotonics,
DOI 10.1007/978-3-642-24172-7, © Springer-Verlag Berlin Heidelberg 2012

MIX
Papier aus verantwortungsvollen Quellen
Paper from responsible sources
FSC® C105338

If you have any concerns about our products,
you can contact us on
ProductSafety@springernature.com

In case Publisher is established outside the EU,
the EU authorized representative is:
**Springer Nature Customer Service Center GmbH
Europaplatz 3, 69115 Heidelberg, Germany**

Printed by Libri Plureos GmbH
in Hamburg, Germany